East-West Scientific Co-operation

Science and Technology Policy of the Baltic States and International Co-operation

NATO ASI Series

Advanced Science Institutes Series

A Series presenting the results of activities sponsored by the NATO Science Committee, which aims at the dissemination of advanced scientific and technological knowledge, with a view to strengthening links between scientific communities.

The Series is published by an international board of publishers in conjunction with the NATO Scientific Affairs Division

A Life Sciences **B Physics**	Plenum Publishing Corporation London and New York
C Mathematical and Physical Sciences **D Behavioural and Social Sciences** **E Applied Sciences**	Kluwer Academic Publishers Dordrecht, Boston and London
F Computer and Systems Sciences **G Ecological Sciences** **H Cell Biology** **I Global Environmental Change**	Springer-Verlag Berlin, Heidelberg, New York, London, Paris and Tokyo

PARTNERSHIP SUB-SERIES

1. Disarmament Technologies	Kluwer Academic Publishers
2. Environment	Springer-Verlag / Kluwer Academic Publishers
3. High Technology	Kluwer Academic Publishers
4. Science and Technology Policy	Kluwer Academic Publishers
5. Computer Networking	Kluwer Academic Publishers

The Partnership Sub-Series incorporates activities undertaken in collaboration with NATO's Cooperation Partners, the countries of the CIS and Central and Eastern Europe, in Priority Areas of concern to those countries.

NATO-PCO-DATA BASE

The electronic index to the NATO ASI Series provides full bibliographical references (with keywords and/or abstracts) to more than 50000 contributions from international scientists published in all sections of the NATO ASI Series.
Access to the NATO-PCO-DATA BASE is possible in two ways:

– via online FILE 128 (NATO-PCO-DATA BASE) hosted by ESRIN, Via Galileo Galilei, I-00044 Frascati, Italy.

– via CD-ROM "NATO-PCO-DATA BASE" with user-friendly retrieval software in English, French and German (© WTV GmbH and DATAWARE Technologies Inc. 1989).

The CD-ROM can be ordered through any member of the Board of Publishers or through NATO-PCO, Overijse, Belgium.

Series 4: Science and Technology Policy – Vol. 15

East-West Scientific Co-operation

Science and Technology Policy of the Baltic States and International Co-operation

edited by

Kazimiera Prunskienė

Lithuanian European Institute,
Vilnius, Lithuania

and

Elmar Altvater

Free University,
Berlin, Germany

Kluwer Academic Publishers

Dordrecht / Boston / London

Published in cooperation with NATO Scientific Affairs Division

Proceedings of the NATO Advanced Research Workshop on
Science and Technology Policy of the Baltic States and International Co-operation
Vilnius, Lithuania
October 3–6, 1996

A C.I.P. Catalogue record for this book is available from the Library of Congress

ISBN 0-7923-4689-0

Published by Kluwer Academic Publishers,
P.O. Box 17, 3300 AA Dordrecht, The Netherlands.

Sold and distributed in the U.S.A. and Canada
by Kluwer Academic Publishers,
101 Philip Drive, Norwell, MA 02061, U.S.A.

In all other countries, sold and distributed
by Kluwer Academic Publishers,
P.O. Box 322, 3300 AH Dordrecht, The Netherlands.

Printed on acid-free paper

Printed in the Netherlands

CONTENTS

List of Contributors

KAZIMIERA PRUNSKIENĖ
Former Prime Minister of the Republic of Lithuania
President of The Lithuanian-European Institute
Vilniaus St. 45-13, 2001 Vilnius, Lithuania

PROF. DR. ELMAR ALTVATER
Free University of Berlin, Department of Political Science
Ihnestr. 21, D-14195 Berlin, Germany

MR. MINDAUGAS STANKEVIÈIUS
Prime Minister of the Republic of Lithuania

MR. V. RUSAKEVICH
Deputy Prime Minister of the Republic of Belorus

PROF. DR. BORISAS MELNIKAS
Vilnius Gediminas Technical University
Saulëtekio Al. 11, 2600 Vilnius, Lithuania

DR. ASTA LAPINSKIENË
Klaipëda University

DR. KLAUS-HEINRICH STANDKE
President of Internationale Akademie Schloss Baruth
Taubchenstrasse 1, D-14163 Berlin (Zehlendorf), Germany

DR. .ARNDT HOPFMANN
Free University of Berlin
Ihnestr. 21, D-14195 Berlin, Germany

LEONID F: ZAIKO
Director of the Institute of Independent Research, NCSI "East - West"
Minsk, Republic of Belarus

V.D. NOZDRATCHOV
President, Jupiter Stock Company
Krestovskij Pr.9, P.O. Box 934,
197042 Saint Petersburg, Russia

ERIKA SCHWEFEL
GESIS
Schiffbauer Damm 19, D-10117 Berlin, Germany

DR. LUTZ MEZ
Free University of Berlin
Schwendener Str. 53, D-14195 Berlin, Germany

DR. NYKOLA BALTAZHI
Counsellor of the Embassy of the Republic of Ukraine in Lithuania

MR. STUART J.D. SCHWARTZSTEIN
US Office of Naval Research Europe
Edison House,223/231 Old Marylebone Road,London NW1, United Kingdom

INTRODUCTION

Science and Technology Policy of the Baltic States and International Co-operation

NATO SCIENTIFIC PROGRAM
OCTOBER 3-6, 1996, VILNIUS

This workshop was attended by approximately 40 of the 70 participants who had applied for inscription and temporarily took part in the workshop's deliberations. The 20 key speakers, which prepared presentations, came from Lithuania (5), Latvia (1), Estonia (1), Belorus (2), Russia (1), Ukraine (1), the USA (2), France (1), and Germany (6). Some of the texts of the presentations were distributed in advance. The working languages of the workshop were Lithuanian, Russian, and English.

The State of Science and Technology in the Baltic Region

The topics discussed were science and technology policy of the Baltic States and other former Soviet Union countries, and co-operation with the West. Some reports gave and overview of the science and technology potential of the concerned states and on science and technology policy performed by the governments after independence more than five years ago. Others attempted to assess science and technology and the potential for future economic and social development. It is a common feature of all former Soviet Union countries that the past is a heavy heritage other to centralisation of the scientific personnel, separation of research from teaching activities at universities, a concentration of certain branches of technology in different parts of the former Soviet Union, so that the newly independent Baltic States dispose on a heavily one-sided technology and science potential. Lithuania, for instance, inherited large parts of the electronics industry of the former Soviet Union. But, since industrial production after independence declined, the necessary linkages between industrial production and the sector of technology and science have been cut, so that it is now rather difficult to revitalise the technological potential. It is also a problem that the financial crises of the newly independent Baltic States does not allow supporting universities and R+D institutions, may they be publicly or privately organised. The speakers had mentioned, that the S+T priorities, formulated by the governments, are not supported enough by the evaluation of the international market and real possibilities for entering into it.

K. Prunskienė and E. Altvater (eds.), East-West Scientific Co-operation, 1–5.

Co-operation among the Baltic States in science and technology is still in the beginning stages. One of the reasons is - as some speakers said - the over-exagerration of the recently acquired national independence. But it was almost common sense that in the area of science and technology policy co-operation must be enhanced not only for the production of a "critical mass" of scientists and scientific output from science and technology institutions. This assessment was given for the natural sciences as well as for social sciences and humanities. Project management is required for co-operation in the area of S+T, as well as for the fostering of industry. Groups of experts from the NATO, Baltic, and Eastern countries are necessary for preparing project studies and investment projects. Their financial interests must also coincide.

International co-operation with the West is still hampered by linguistic problems. Therefore, it has been recommended that the West support language studies in the Baltic region in order to increase the capabilities of active participation in the international scientific community. The enlargement of linguistic capabilities is also necessary to make use of the potential of electronic media, in particular, electronic communications such as the Internet. There are already Net-servers installed in the Baltic States, but the distribution and participation in electronic communications is still in the early stages.

A critical issue of international co-operation between the West and the Baltic States is finance. Due to the budgetary constraints in the Baltic States and considering financial cuts in the West, the financing of international co-operation in science and technology is difficult. It has been mentioned that in science, public funds have been requested, but in technological development, and technology transfer, private institutions must take the initiative. The government policy has been criticised for contributing only to science institutions and not for concrete S+T studies and projects. It does not support the orientation towards the "know how" market needs.

Analytical Issues

The reports on the state of science and technology policy and the assessment of co-operation in science and technology policy in the Baltic region, as well as between the Baltic States and the West were followed by analytical issues, raised in some of the presentations submitted to the workshop. Some presentations focused on specific questions, such as in the sectors of energy policy, electrical equipment and electronics, industries which are of utmost importance for most of the Baltic States. In the case of energy policy, environmental questions were also discussed. The urgency to find solutions to the energy problem has been stressed not only by the paper giver but also by the presenter, but also by many of the participants, since energy is still a bottle-neck sector in the Baltic region. A very actual problem is the integration of the Baltic States' energy sectors into the Western Energy System, as well as its restructuring, and the use and lowering the price of local resources.

Another issue was the role of science and technology transfer for international competitiveness of industries situated in the Baltic area. In order to increase competitiveness it is not sufficient to support and foster single firms. Access to high technology is only one prerequisite of improved competitiveness. It is necessary to mobilise the contributions of all policy areas for an improvement of science and technology. Science and technology are elements of a many-fold system of factors responsible for the competitiveness of nations. It has also been pointed out that science and technology cannot be viewed without taking social issues into account, since social acceptance as well as networks between different institutions of a given society are important for scientific and technological progress.

The transfer of technology from technology givers to technology receivers was also an important issue. Which channel is the most efficient? Technology transfer via trade relations, via direct foreign investment, via migration? The answers have been rather controversial. The question of the consequences of the brain drain suffered by many of the former Soviet Union countries was of utmost interest to the participants from the Baltics. The causes of the brain drain seem to be rather simple: low wages for valuable science and technology personnel, compared to the remuneration paid in private industry, especially in the West. Brain drain could be a future jeopardy for the systematic character of science and technology development. Therefore, it has been argued, everything possible should be done in order to install an incentive system which would hinder the brain drain from the East to the West.

The competition is growing between the West and Southeast Asia, with the latter's products pushing their way into previously Western controlled markets. The Western countries are trying to import production, not products (which is important for securing jobs, increasing budget incomes, and the general economic potential). The opinion of scientists and electrical, as well as radiotechnological, industry leaders in Lithuania and other regional countries is that that kind of policy direction is necessary in their countries also. If the Western countries opt to only export products into a region, without developing the local region's production, then the market is not being activated - this was also the opinion of some participants from the West. Many examples were given of how the West receives the potential, from Lithuania and other regional countries, almost at no cost and then, on that basis, improves the new technologies and then exports them back into the region.

This argument has been underlined by participants stressing an aspect of science migration which normally is not discussed, that is the possible use of scientists migrating from the East to other countries and of science (and scientific results) being available for a small amount of money in the former Soviet Union by criminal subjects, by terrorists or by governments supporting terrorists. This danger could also be used as an argument for political regulation of the brain drain.

Recommendations

During the seminar, many recommendations were made with regard to economic policy, social policy and S+T policy. In the area of economic policy, the participants stressed the necessity of political support for the initiation of a "virtuous circle", which in the case of the Baltic States, can only be established with great difficulty. The circle consists of four basic elements: minimum market size, production requirements, technology and science. Production will follow the market, only if the market size is large enough. An indigenous technology potential will be required only when there is industrial production in the region, and science will flourish only when institutions providing technology have a future perspective. However, the market size of the Baltic States is very small, and even smaller than it should be. Therefore, one recommendation was made to intensify regional co-operation in economic policy, and to foster co-operation with Russia, in the hope to gain access to the Russian market. For regional integration sectors in the Baltic area, financial stabilisation in an unavoidable prerequisite. Supporting financial stabilisation programs should, therefore, be one of the objectives of co-operation with the West.

In general politics, it has been underlined the enhancement of science and technology needs the political will of the political elite in order to prepare the three Baltic States more so than it has been during the first years after independence. Very often it has been mentioned that the Baltic States may be successful insofar as they are able to find "niches" within the world market, where they are competitive. But the task of finding niches cannot be the work of the private sector alone. There must be some political support by the governments, and there must be a social discourse on the future aspects of science and technology development conforming to global developments. An assessment of the Baltic States' science and technology industry potential was offered, with which a concrete offer could be made for co-operation (joint project) in the market - how the Baltics fit into the world market, and for international science, technology, as well as industrial and economic integration. The discussions showed what the actual problems of co-operation ethics and trust were from all sides (West - Baltic - East). The West sometimes devalues the potential of new partners from the East and Baltic countries, they would like to use it at a low cost, and they do not trust the ability for independent work or possibilities for their use. Whereas, scientists and industrialists from the Baltic, and other developing countries, after experiencing this devaluation, psychologically begin to distrust the Western partners and the transfer of new technologies and their development in joint investment project trends.

Therefore, in the areas of education policy and social policy, open-mindedness of the people must be supported by all actors in the society. One of the main prerequisites is the strengthening of linguistic capabilities, with particular emphasis on the English language, and more contact with people in other parts of the world. Here, western institutions could rather inexpensively assist the scientific development of the Baltic States.

Activities of Continuation

The original idea, upon which this seminar was organised - to move from the discussion of concrete ideas to the study stage and investment projection - was upheld. This offer was addressed to the NATO Scientific Division with the belief that an application would be written for the possible financing of a future program, which concurs with the requirements of NATO.
At the end of the workshop, there was also a discussion on possible continuation. Two themes have been singled out as to be of greatest importance and interest to the Baltic societies. The first is the question of legal market framework, i.e. market regulation in the case of market failures, intellectual property rights, etc. The second is environmental policy in the Baltic region and co-operation with the West. Environmental policy of course, has scientific and technological aspects, insofar as efficiency issues are concerned. But, it has much more to do with individual and social behaviour, institutional regulations, etc.

One of the most positive aspects of the workshop was the establishment of rather durable contacts, between scientists within the region and with Western partners, on different issues. So, the workshop was a fruitful contribution to the enforcement of the Baltic scientific community and to the East-West collaboration on Science and Technology issues.

Acknowledgements

The organisers of the seminar and the editors of this publication would like to extend sincere thanks to all speakers and authors, who agreed to publish their papers in this book. Special thanks to Dr. A. Jubier, of the NATO Scientific and Environmental Affairs Division, for his initiative and support regarding the seminar and the publication of materials.

Co-Directors
K. Prunskienė
E. Altvater

THE MECHANISMS OF REGIONAL AND INTERNATIONAL CO-OPERATION IN THE SPHERES OF SCIENCE AND TECHNOLOGY

KAZIMIERA PRUNSKIENĖ
Lithuanian - European Institute

1. The new situation in the preparation of specialists.

One of the most important reform areas in the Baltic countries is the creation of a European system for science, studies and education. The renunciation of the former Soviet instruction and studies order - which is a somewhat easier task than the successive creation of a new system. Those Soviet instruction system principles were quickly disassembled, such as those like:

- centralised and unified instructional program preparation, and the planning of the number of school and university teachers
- mandatory attendance of lectures in universities
- graduate job placement guarantees, and centralised distribution
- financing of studies unrelated to results or effectiveness.

The qualitative restructuring of instruction and instruction methods according to European and world levels, and the formation of teacher, lecturer, professor, science personnel, the qualifications suiting them, and student independence and motivation is a long-term and major effort requiring process.

With the changing of the economic system, the professional orientation changes quickly as well:

- many clever youths, especially during the first year of independence, poured into the business sphere, including the black market, where many became rich quickly, but also lost not only their song, but their life
- most orientation was directed towards the specialisation with good perspective incomes; at that time, the income differentiation reflected the transitional time-period conditions and their instability.

The biggest competition is between entering management, business administration, architecture, psychology, social psychology and pedagogy, German philology, and English philology (see table 1),
whereas the least competition exists entering:
chemistry, physics, geography, many engineering specialities, medicine (which was most popular during Soviet times), and Russian philology. The relation of competition between German philology and Russian philology is 7:1, respectively.

K. Prunskienė and E. Altvater (eds.), East-West Scientific Co-operation, 7–14.

The average competition among higher schools varies from 1.6 - the Lithuanian Veterinary Academy - to 3.9 - Vytautas High University (Kaunas). This is mostly related to the studies content of specialisation. The Vytautas the Great University still the most competitive because it was established during independence together with foreign professional overseeing from the USA. This is the most modern university in which about 500 starting students are accepted every year (Vilnius University - 3 times that number).

One of the peculiarities of the current situation is that the *structure of the specialist market is not yet formed*, and its perspectives are difficult to predict. Most of the earlier high qualifications for engineers susceptible to industry branches such as, for example, electronics and radiotechnics, production of machine-tools and attachments now barely use 30% of their earlier capacity. Most of the highly qualified specialists have found other jobs or have become unemployed.

The state does not centrally plan specialist training, the higher schools decide the number of accepted students per specialisation on their own. However, analysing their structure, it is obvious that this process is forming a long-term directional evaluation changes in the specialists market. This is objectively complicated, because the actions influencing the specialists market are not easily susceptible to prediction - having in mind the chances for industrial and other economic branch development. For this reason, disproportion of a new nature could appear in the future, between concrete professional and specialist supply and demand and drastic changes in the former, as well as in competitiveness.

One of the most important problems implementing the educational and science reform for integration into Europe is the lecturers' *scientific qualification level*, and its modernisation. It is natural that this does not occur during the course of one year. Preparation for higher qualifications for specialists require corresponding competence of the professorship. Some of the universities, unfortunately even now, have not found a solution to this question, because:

- the brightest professors and lecturers went off to work in better paying fields (consultants in foreign companies, establishing consulting firms, working as experts with PHARE and other programmes and so on)
- the state financing of universities and studies centres is very limited, and so is the average wage level in budget establishments which does not allow it to attract strong specialists, and some of which go lecture as community service (like a hobby)
- average wages in Lithuania for science and instruction systems barely make up about 180 USD/month, and for professors - 450 USD/month - the average wage in Lithuania is about 200 USD/month
- lecturing and scientific studies work is co-ordinated, due to limited orders from economic structures and state, has still not received a larger scale, which especially would be actual not only in wage levels, but also in qualification growth, and in the sense of lecture quality.

I would like to note one specific problem in the area of science and scientific - technological co-operation that is general studies (Projekstudie) according to international institutions: EU (PHARE and other), World Bank and other programmes. Until now, it is still elucidated to whether our local experts can aspire to high payments, close (even if not equal) to those of their colleagues - experts - from the West, working in the same structure when there is a huge 20-50 times difference, that is our experts earn 20-30 ECU per work day, which is similar to the average earnings of professors in our local universities.

Differentiating wages between Western and our local experts creates not only an unhealthy psychological co-operational atmosphere worse than an older brother's "phenomenon from the East," but also we lose in the pragmatic respect: the best specialists lose interest in working those projects, choosing private business or business consultation. The possibility doesn't even form for permanent project management which would be able guarantee the continuance of the science chain of ideas - concepts formulated during the study stage until the implementation of investment projects. Germany's business structure representatives support the view of this question which I have presented, understanding how important it is to relate studies to realistic possibilities of not having to put the results away in a drawer, but to use them in practice.

Therefore, I would like to ask the German government to support of partnership, setting non-discriminatory co-operation principles in the EU and German foundation financed projects.

For the general higher schools, as with their independent science and studies centres , the *level of financing* is low and dependent on the economy and the abilities of the state budget. Preparation for NATO membership requires more funds to be allocated to defence, which in 1997 will increase by 85%, at a time when science investments have increased only 15%, and the inflation being 13%. This allows for the doubt of this policy's well-foundedness, understanding that national security is no less than before dependent upon science - technology and economic development, social welfare and stability.

New opportunities in science and higher learning is based on co-operation with Western higher schools, activating general scientific studies, probationary work, academic exchanges, study programmes, common seminars and practice (training). In this sense, the state of co-operation according to separate schools is very unequal. There are too few ties between technical and pedagogical universities. The spread of direct and continuous ties would allow very concrete opportunities for quality changes.

Highest and lowest competitiveness among higher schools in 1996

Higher school	Highest competitiveness Study programmes	competitiveness	Lowest competitiveness Study programmes	competitiveness	Average competitiveness
1	2	3	4	5	6
KTU	Management Architecture	10.7 8.5	Electronics	1.9	2.9

	Business Admin.	8.3			
	Tech. and Design	4.0			
	Informatics	3.6			
KU	German philology	7.2	Music	1.4	3.3
	Lith. phil. and	7.1	Engineering prog.	2.3	
	Swedish		Choreography	2.5	
	Practical	7.0	Children's' pedagogy	3.0	
	mathematics	6.1			
	Economics				
	(Business admin.)				
VPU	English	5.8	Chemistry	1.2	3.0
			Belorussian	1.3	
VTU	Institutional	7.1	Fire protection	1.6	3.0
	admin.	6.1	Bio-mechanics	1.8	
	Business admin.	4.8	Automatics	1.9	
	Architecture		Radio-electronics	1.9	
VU	English philology	5.6	Chemistry	1.0	3.3
	in Kaunas		Statistics	1.0	
	German philology	5.5	Physics	1.1	
	Psychology	4.8	Russian philology	1.1	
	Journalism	4.7	Clinical Biology	1.1	
	English philology	4.0	Mathematics	1.2	
			Geology	1.3	
			Geography	1.4	
VDU	Psychology	8.2	Physics	1.3	3.9
	English lang. and literature	7.2			
	Lith. lang. and lit.	7.0			
KMA	Nursing (after higher med.)	4.2	Nursing (after secondary school)	1.9	2.5
	Community health	3.6	Pharmacy	2.2	
	Dentistry	3.0	Medicine	2.2	
LKA					1.4
LMA	Singing specialisation	3.4	Ethnic musicology		
			Composition		
			Musicology		
LPA					2.6
LVA	Veterinary	2.0			1.6
LZUA	**Day studies:**		Engineering studies programmes	1.6	2.0
	Home econ. tech.	2.8			
	Agriculture	2.6			
	Forestry	2.5			
	Correspondence courses:				

	Agric. accounting and financing Agric. economics Forestry	4.4 3.8 3.3			
VDA					3.2
LKKI	Tourism and athletic administration Kinetic therapy	4.3 4.3	Health nurturing	2.3	3.4
SPI	Social psychology and pedagogy English Special pedagogy and body culture Art and design Elementary teaching and body culture	8.9 5.0 4.8 3.2 2.9	Physics and math Lith. lang and lit. Elementary teaching and art Elementary teaching and pre-school nurturing	1.4 1.6 1.6 1.7	1.8

Full Names of Higher Schools by Initials

KTU	Kaunas Technological University
KU	Klaipeda University
VPU	Vilnius Pedagogical University
VTU	Vilnius Gediminas Technological University
VU	Vilnius University
VDU	University of Vytautas the Great (Kaunas)
KMA	Kaunas Medical Academy
LKA	Lithuanian Military Academy
LMA	Lithuanian Music Academy
LPA	Lithuanian Police Academy
LVA	Lithuanian Veterinary Academy
LZUA	Lithuanian Agricultural Academy
VDA	Vilnius Art Academy
LKKI	Kaunas Institute for Body Culture
SPI	Siauliai Pedagogical Institute

2. Equal partnership in co-operation: urgency and difficulties.

The Baltic countries, participating in the programme "Partnership for Peace," must first of all guarantee security and peace for their own region, which depends not only on power structures, but mostly on political and social stabilisation, scientific potential, and economic revival. It is impossible to overcome the recreation of jobs, the recession of wages of residents and the living standard, and the stagnation of intellectualism only through the creation of small new economic subjects. For science institutions and larger industry to recover, new long-term co-operative ties are necessary. I n science branches, industrial co-operation moves parallel to co-operation in the sphere of science and technology.

For the essential co-operation inactivity and its reasons, as well as the balance of relations, it is necessary to consolidate the Baltic region's scientific and industrial potential, and the development of its evaluation in the new market context, and inter-regional co-ordination. This could be a very important precondition for nurturing initiative in the view of the Western partners, and in some areas, it will allow for the procurement of enough modern critical mass and its disposal will become the half for the actual choosing partners, including those of the West. The role of full partners and the goal of partnership balance is not a matter of professional ambition or politics. It is of vital importance for New countries in the social and economic sense. Having prospective project offers and initiative, competitions can be announced for full-fledged partnership roles, themselves inviting Western companies for shared project financing and implementation, not giving up, but sharing local Baltic region, Russian, and other country markets.

Up until now, however, inter-regional science and technology co-operation is weak and poorly co-ordinated. General projects move mostly through the EU and other financial structures and limits projects to infrastructure branches. Industry is almost never involved in these studies and investment projects. The reason for this is that without end product "know how" and mastery of financial markets of Western partners: those in the roles of clients, investors, project managers, and strategic investors the same regional science and studies organisational and production companies have very limited possibilities to move from ideas to creation and end product production.

The EU "PHARE" and other shared programme studies and projects cannot force the science and technology potential into motion. The science and engineering personnel in the Baltic and other reforming countries which are pulled into shared projects with foreign experts are being implemented on an unequal basis. They themselves cannot be either direct competition participants (only expenses or sub-contractors), nor be trained on an equal basis. Thus, as was already mentioned, the most highly qualified specialists are re-orienting themselves to better paying jobs, including business (often in areas where less qualifications are required). At the same time, there appears the problems with co-operation ethics and trust. The West, depreciating our potential which is actually more valuable, do not trust and would like to purchase it for the least expensive price. Our countries' scientists and producers, experiencing this depreciation psychologically disqualify and do not trust the Western partners' offers and good will. Also, shared projects are not only connecting knowledge and experience, but it is also the synergy effects of the shared effort, in which psychological harmony, a creative atmosphere, effective functioning, and the distribution of funds is necessary.

The formation of project offers and their marketing in the international market, seeking long-term science, studies, and complex production co-operation, could have success only with the existence of common interests and activity. So far, the most active Western partners , with our businessmen, are working:

- trading with Russia and the other CIS countries
- ordering products requiring low scientific qualification from our companies
- importing Western products into our market.

These types of co-operation do not stimulate the unity into science, do not form satisfy the business need for orders from scientists and inventors, which lose their financial sources.

With the crumbling of the Soviet Union and socialist ties with the former industry specialisation and co-operation, most of the science-studies establishments and industrial enterprises, especially electronics, lost their former markets and continue to use their former capacity, only partially and sporadically, at only 20 - 30%. This situation is considered good only due to exceptions. A portion of those branches with scientific potential cross over to other commercial economic branches or have "drained" out into foreign countries. The remaining potential: qualified specialists, technological, management, and marketing require serious modernisation and integration into world science, technological, and market systems. The more complicated branch of this lies in the fact that more efforts are required, implementation of new projects and qualifications.

If partners in the markets of crude oil products, metal or food products find one another without any extraordinary difficulties, they could easily change partners and here, direct sales-purchase transaction is relevant, when "know how," technology, complex apparatus and equipment - as an end product - producers work together to prepare the necessary marketing and project management for the product. The problem is that until now, our enterprises most often have not independently mastered their production and possible, or offered, co-operation project marketing. They remain in the passive role, waiting for an abstract partner who must enter as a foreign investor or client and solve the problems. Consulting enterprises are called upon infrequently or for superficial reasons.

3. Mechanisms for new opportunities.

For the results of the activation of science and its demand, the financing order must be changed, for which the Lithuanian government policies are worthy of blame, when the budget funds are allocated to science are directed toward not actual science and technological modernisation and projection institutions. The international market needs are little studied and valued. This financial structuring does not stimulate the actual scientists to orient themselves into modern market projects, it upholds the "vegetarian" mood, and does not stimulate attempts to participate in competitions for orders and production. The mentioned government activation necessity, representing EU and other international institution, as well as strengthening equal conditions in the implementation of science and studies projects together with Western partners, and liquidating discriminatory conditions for Baltic and other countries' experts. This is actual not only for the scientists and specialists for motives of financing and psychology, but for general state interests, attempting to not lose science personnel and that studies sequentially move towards practical application without losing project managers and consultants.

The Lithuanian-European Institute, and the NATO Scientific Division financed, together with Prof. Dr. Elmar Altvater (Free University of Berlin), have already prepared two international Workshops devoted to science and technology policy, industrial conversion and international co-operation in these areas. There is no doubt as to the seriousness of these topics, which is present in the active discussions. We believe, that in our move away from strategic questions would be proper in the move towards concrete studies and investment projects. The electronics and radio-electronics industry, whose successes in the Baltic region and specialist qualification had greatly out-grown the science and technological levels of the Soviets, could be that priority co-operative object, which could join the forces of the NATO countries and the Baltic countries. We address this offer first to the NATO Scientific Division, in the hopes of receiving their support and to form, in the near future, a general expert - specialist group to begin studies.

The LEI, in it's own right, pledges to collect and present the necessary information about Lithuania's (according to possibilities - even the Baltic region) electronics and radio-electronics industry's current potential, and new possibilities and offers for co-operation projects. This should surely assist the NATO Scientific Division in their decision about a projective studies profile (area). From the concept of the idea, to projective studies, to implementation is a long road. We should not stop half-way along the road, but go the full distance to realisation.

One other concrete suggestion is to establish a small university based upon the Western (e.g. Germany's) model, to whose thanks we could prepare management, financing, banking, and other courses related to business import-export, investment, technological modernisation of production, and other similarly related specialisations. This has been a long-hoped for idea, which could be, from Lithuania's side, be implemented with much enthusiasm. The university would be an important contribution to quality integration and concrete co-operation, as well as for the development of results. We would be grateful for this kind of concrete support from our Western partners for science and economic development.

INTERNATIONAL COMPETITIVENESS AND SCIENCE AND TECHNOLOGY TRANSFER

ELMAR ALTVATER[1]
Free University of Berlin

In history, there are not many positive examples of successful science and technology transfer from one economic and social culture („technology-giver") to another („technology-taker"). Most attempts to transfer science and technology failed, since transfer is a very comprehensive process and means more than just the transfer of recipes, blue-prints and sometimes of „brain" („brain drain") from one country to another. After the Second World War, for instance, `science migration` from Germany to Latin America (especially to Argentina and Brazil) or to the Soviet Union was not very successful if measured with the expectations of positive input of science into the technologically nurtured growth process in these countries. There were partly positive results with regard to the United States or England, where foreign scientists have been integrated into an already existing scientific culture with highly formalised organisational structures and an already existing highly developed industrial background.

These experiences in the 60s were an incentive for the Hungarian economist Franz Jįnossy („Am Ende der Wirtschaftswunder", Frankfurt 1966) to develop an alternative economic interpretation of growth processes compared with „mainstream"-economics. Quite recently, his approach has been „re-invented" by the much discussed „new growth theory". He found out that the causes of the „economic miracles" (in plural) after the Second World War in most European countries (Western as well as Eastern Europe) which have been affected by the destructiveness of the war was the structure of qualification of manpower and its very inertial development. Capital investment, without any doubt, is important, but not the decisive pulling (or pushing) factor of economic growth (as assumed in most neo-classical growth models). This analysis corresponded rather well with the later findings of Mac Kendrick or Denison and others, detecting a „third factor" beyond capital and labour (or land) responsible for economic growth. They called this factor „technological progress". But technological progress is much more than

[1] Free University of Berlin, Department of Political Science, Ihnestrasse 21, 14195 Berlin, Tel: +49-30-8384965, Fax: +49-30-8386347

K. Prunskienė and E. Altvater (eds.), East-West Scientific Co-operation, 15–27.

just technical inventions and blue-prints, it comprises - as already mentioned - qualification of manpower, demand factors (effective income), supply of energy and matter, externalisation possibilities of pollutants, risk management, participation and engagement of people (see graph no. 1). Economic dynamics, therefore, depend on more than mere technological and scientific inputs; but without any doubt, they are one of the most important factors.

Graph No 1

Growth and development as a process of domination of energy- and matter flows, of risk management and of distribution; Systemic character of the "growth machine"

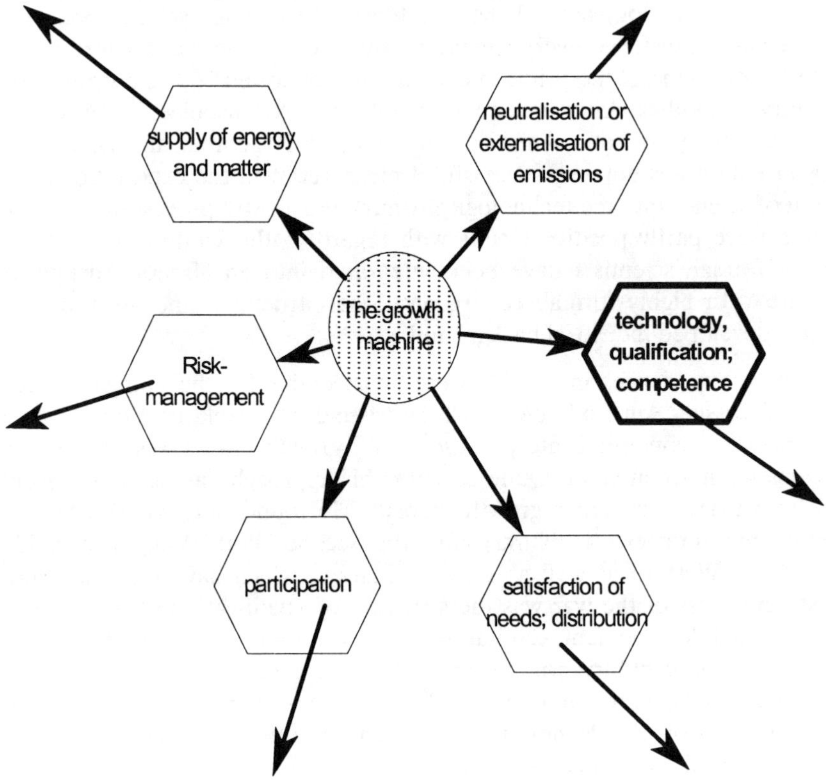

Jịnossy concluded that technological progress, of course, is a very important factor of economic development, but only as far as it is not too far ahead of the slowly and inertial developing structure of qualification of manpower. Scientific progress, therefore, is a fore-running and pulling factor, but the distance to the inertial qualification structure must not be too big. This, of course, is an efficient limiting factor of endogenous technical progress as well as of the exogenous implantation of transferred science and technology. The „absorptive capacity“ in a given country is of utmost importance for the efficiency of scientific and technological inputs in the economy („growth machine“) and the spill-overs.

Mostly, Joseph A. Schumpeter is quoted when the role of technical innovations are on the agenda of theoretical and political discourses. Schumpeter's argument is partly different from the approach chosen by Jịnossy. He stresses the other side of the coin by saying that technical innovations, *firstly,* are produced endogenously in the course of the economic process. This is a basically classical argument shared by Adam Smith or Karl Marx. *Secondly,* innovations are an expression of „creative destruction“ in the course of social progress and, therefore, not beneficial to all actors in a given society. There are winners and losers of technical progress. *Thirdly,* innovations are the result of entrepreneurial decisions following the motive of maximising profits. The technological path chosen, thus, has a „capitalist bias“. This means that other motives in technological and scientific development, such as social adequacy or environmental protection in a market economy can only play a role as long as they do not contradict the profit motive of micro-economically rational subjects. *Fourthly,* innovations occur the faster and the more massive the more highly developed science and technology („big science“) and qualifications are available in a given society. Obviously, the fourth element can be modified by science and technology transfer.

By means of technology and science transfer, a country can rely on science and technology innovations developed abroad in another „scientific community“. Of course, tapping science and technology resources of other countries is only possible as far as results of science and technology in principle are a public good. In the reality of a Schumpeterian world ruled by the GATT- and WTO-regulations (trade-related property rights and trade-related investment measures, intellectual property rights etc.), the access to science and technology is only possible under very restrictive conditions. These conditions are the more restrictive the more important science and technology are for the competitiveness of single corporations or whole national economies in global competition. Therefore, the constraints are usually especially high in „high-tech“, i.e. in the spheres of information, communication, biotechnology and genetechnology etc. As far as science and technology are subject to the „exclusion principle“ of patents and licenses, they are monopolised by private trans-national corporations. For developing countries as well as for transformation countries, the access to high technology and science, as long as it is monopolised by TNCs, is thus rather limited. Results of science and technology, although in principle a public good accessible for everybody, in the globalised world, under the regime of patents and licenses, they are private goods. As commodities, their production and distribution follows the market (price) mechanisms and therefore private interests of trans-national corporations.

Moreover, globalisation of the economy also means globalisation of norms and standards. Therefore, the power of influencing or determining the standardisation of products and processes of production is an important factor of local, regional and national competitiveness. It is important for the position of TNCs on the market whether technical norms and standards are defined in Europe, in North America or in Japan, because market access in most cases depends on the compatibility of devices with effective standards, beginning with the plugs of electrical devices and not ending with environmental norms of car emissions. Standardisation (the ISO norms), however, has a decisive effect on technological trajectories. They may exclude some trajectories chosen by countries or firms in the past which, following the definition of a viable path compatible with globally defined norms and standards, proves to be a *cul-de-sac*.

Many experiences in the case of developing countries show that private enterprises are only willing to transfer high tech as long as they have the guarantee that the use of that technology remains under their control. So it is not certain that a science and technology transfer can lead to the introduction and integration of new technologies into the endogenous economic circle of a given country. This is comparable to the technology transfer in the arms industry which only takes place if it can be controlled by the transferring country. The reason for this behaviour of private corporations is rather simple: transferring technology without control only would nurture new competitors. Therefore, transfer of technology preferentially takes place in those areas which are not decisive for the competitive position of transferring firms or countries. As a consequence, science and technology transfer is more likely to take place in the case of relatively old and therefore already wide-spread technologies or in those cases where existing technologies can rather simply be substituted for new technologies which are not transferred.

These are important factors for the „technology giver". Concerning the „technology taker", there are also obstacles for technology transfer. The necessary embeddedness of science and technology in a given qualification structure of manpower in a country has already been mentioned. Additional factors which are responsible for the „absorptive potentiality" of a country for transferred science and technology are: (1) a sufficient degree of „systemic intelligence" in microeconomics and macroeconomics institutions and organisations, (2) administrative competence, networks of contacts between the spheres of science production and the application in industry, and - see graph no. 1 - (3) capability of risk management which very often is underestimated as a technology-relevant factor. Only with these *non-market factors* of technological and science absorption, it is possible to *adapt, to imitate, to develop and implement* technological innovations imported from abroad. There are many examples all over the world that technological spill-overs do not sufficiently occur, even in a country characterised by a certain homogeneity of the economic and social structure. In the case of two countries being rather heterogeneous in terms of economy, society, culture and with regard to the level of economic and technological development, transfer of technologies and science is rather likely to fail. The consequence is rather simple: technology and science transfer only make sense if they are not an isolated measure of single firms but a political project pursued by firms, governments and social institutions or organisations. It has to become an integral part of „industrial policy“, comprising market regulations as well as political interventions by the state and the mobilisation of social networks.

Scientific and technological co-operation on an international level, therefore, must be accompanied by an adequate qualification and science policy of the nation state and regional political entities, as well as by the creation of a favourable and stimulating social climate for scientific and technical progress, including an economic policy fostering macro-economic stability. This seems to be rather obvious, and in many cases, this recognition has been transformed to a pure policy of fostering „social acceptance“ of technical development. It is rather likely that the combination of markets, politics and social networks does not work in this simple way. Scientific and technological progress only stand a chance when social interests are involved - at least in a democratic society.

There are still international factors to be taken into account. A glance back into the last decade of the 80s in the developing world reveals that macro-economic stability depends on global and not only on national parameters. The so-called debt crises of developing countries diminished the import capacity of these countries as well as the interests of trans-national corporations to engage in long-term direct investments in these countries. Therefore, these countries today are technologically lagging behind the industrialised countries to a greater extent than before, although UNIDO, at its Lima conference in 1975, predicted a catching-up process of the less developed countries with the industrialised world by the year 2000. This sad experience paradoxically is reflected for instance in the „White Book“ of the Commission of the European Union of 1993 on „Growth, Competitiveness, Employment“ and the „Challenges of the Present and Pathways into the 21st Century“. Developing countries are only mentioned as possible markets for European firms, not as possible competitors with technological competence of their own on European markets. This is quite different with regard to the competitors in North America or East and South East Asia which are taken seriously. With regard to Central and Eastern Europe this might be different because scientific and technological competence as well as the qualification structure of manpower are already highly developed, at least much more developed than in the case of most „third world“ countries. Moreover, after the revolutionary transition from 1989, it is possible to build plants on the „green meadow“ in an „extraordinary market-friendly environment“, as the IMF characterises CEE countries. Another reason for a special treatment towards Central and Eastern Europe is geographical and cultural proximity. This is a very important factor for scientific and technological co-operation which must not be underestimated.

Graph No 2
Aspects of Technology Transfers

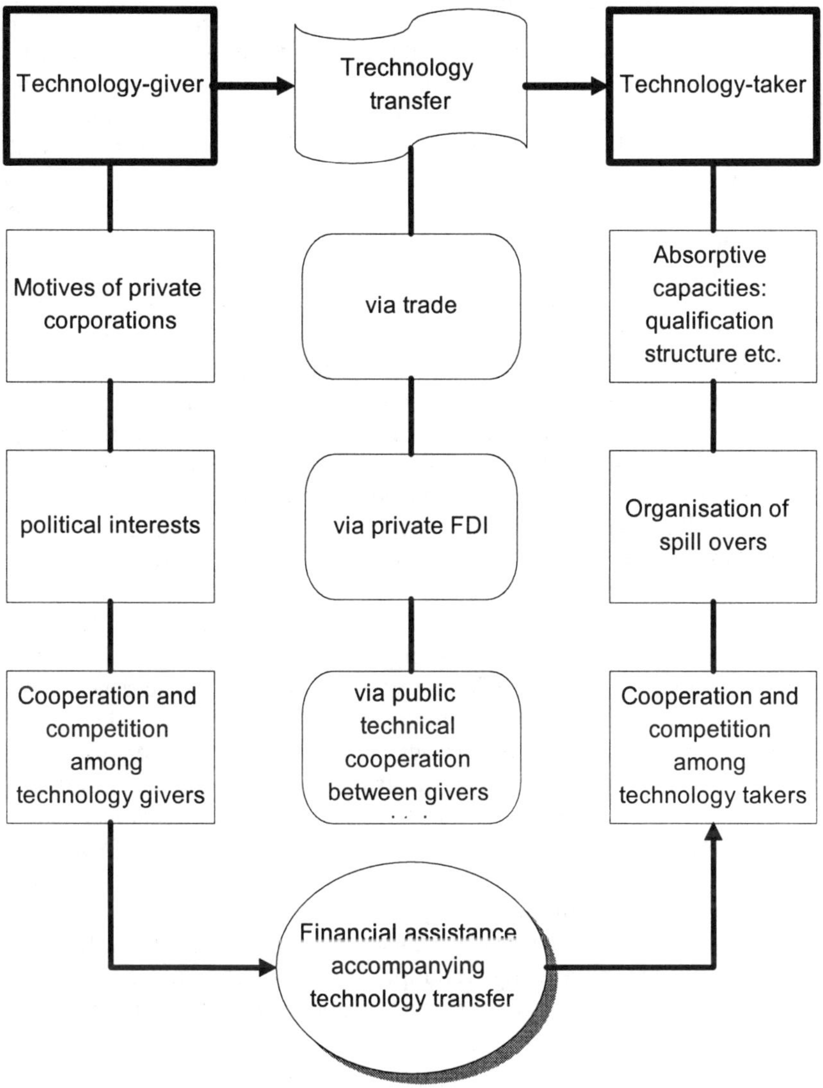

Technology transfer regularly takes place in three different ways (see graph 2). The *first* is that of commodities in which given technologies and scientific results are incorporated. It is quite obvious that this kind of technology transfer is more likely to take place in the case of investment goods compared with consumer goods, and that it is also more important in the case of manufactured goods than in the case of raw materials. The technological content of commodities, therefore, is decisive for technology transfer.

The *second* path is that of foreign direct investments undertaken by trans-national corporations. Investment decisions follow the motives mentioned above. Therefore, this way is very important but it should not be overestimated concerning the spill-over effects of technologies, transferred by means of direct investments. A glance at the regional distribution of foreign direct investments and on the distribution among industries clearly shows that the expectations in technology transfer should not be too great. In the case of Germany, nearly 90% of direct investments are applied in EU-countries (around 50%) and other industrialised countries (40%). Only 10% are going to developing countries and reform countries in Central and Eastern Europe. Their share in German foreign investment only amounts to about 2%. The interest of „technology givers", as far as private firms are concerned, is apparently rather low with regard to less developed countries. Technological and scientific co-operation is more attractive with similar partners than with those still needing technological development. Most technological co-operation among trans-national corporations therefore takes place among TNCs: within the US, transfer takes place between domestic TNCs or between domestic TNCs and Japanese or Western European TNCs, as Archibugi/Michie (1994) demonstrated. Again it is quite clear that the reason is connected to the arguments mentioned above: „It remains the case that without an innovative local society supported by adequate professional organisations and public institutions there will be no innovative milieu. And without an innovative milieu the development of high technology industries will contribute to regional development only within the heavy constraints set by the business cycles of industries that are likely to be highly volatile. There will be no possibility of truly indigenous growth and thus no escape from the state of dependency on another region, another region's companies and another region's innovative individuals" (Castells/Hall, 1994, p. 235, in: Altvater/Mahnkopf, 1996, p. 275). Therefore, the conclusion on the relevance with regard to science and technology transfer of this second path is rather disappointing: It is the path which links technologically and in terms of science structure and culture already developed regions and countries; it is less apt to channel technical and scientific co-operation between countries of different technological levels.

The reason, again, has to do with the systemic character of science and technology. In the case of the United States, it has been shown (by Marcus and Segal xxx; Altvater 1992) that science and technology have systemic characteristics as an economic growth factor. The application of high technology as well as scientific progress only occurs in a complex environment of social and cultural institutions, of professional organisations and alliances between industry, public administration and science and technology institutions, of co-operative structures among scientists, engineers and technicians. But since the scientific and technical community belongs to society as such, the persons performing scientific and technical projects have to be integrated in social institutions and activities. Moreover, the other side of co-operation in scientific and technological progress is competition; without competition, co-operative arrangements lead to stagnation. The environment of technological and scientific progress, therefore, is rather contradictory, but it requires a minimal degree of systemic coherence in order to fulfil its function of producing and fostering technical and scientific innovations.

The *third* path of technology transfer is that of migration. But this path normally is counterproductive. As already mentioned above, science migration from Germany to the Soviet Union or Latin America following the Second World War was not very successful for the recipient countries, because the absorptive structure was not efficient and sufficient. Since 1989, this may be different, because scientists prefer migrating from East to West rather than vice versa. Therefore, in a situation where different development levels reflect substantial wage differentials, a brain drain from less developed countries into more advanced developed countries is more likely than vice versa. Over a longer period of time, this brain drain may seriously hurt the countries the migrants come from, whereas the positive effect in the absorbing country is not very remarkable in most cases.

The *fourth* path of technology transfer is official technical co-operation. Technical co-operation is the ordinary way of technology transfer in the larger policy area of development assistance or foreign aid. But normally, technical co-operation is not organised to support an increase of competitiveness of the recipient country on global markets. Its objective in most cases is alleviation of poverty, improvement of urban and rural infrastructure, basic qualification measures to enhance professional activities or environmental protection. Therefore, public technical co-operation plays a considerable role in rural regions and for urban small and medium-sized industries, in measures concerning social and material infrastructure, but not in those industries which participate in world markets and therefore in high tech competition. The main objective of technical co-operation therefore is circumscribed by the word „empowerment“ and not targeted to increase competitiveness.

Only in very few cases, technical co-operation takes place in modern high tech industries. The nuclear contract between Brazil and Germany of 1975, for instance, which also might be interpreted as a warning for the pitfalls of this path of technology co-operation. More than 20 years since the nuclear contract between Germany and Brazil, the atomic power plant of *Angra dos Reis* does not produce electrical energy at all, although 2.5 billion US dollars have been invested. This is one of the most expensive investment ruins of Brazil. It demonstrates once again that nuclear power is a technological *cul-de-sac*.

A very interesting case of public technical co-operation is the transfer of technology and scientific know-how as it is stated in the climate convention of Rio de Janeiro (1992) and Berlin (1994). In article 4, 5 technology transfer is also mentioned. It is the common responsibility for the global climate that led to the regulation of article 4. Technological transfer is tied to financial transfer from developed to less developed countries. It is also supposed to improve technologically indigenous achievements and the social and material infrastructure. The expectation of this kind of technology transfer is remarkable reduction of emissions relevant for the climate of planet Earth. A special case of transfer of environmental technologies is the establishment of „joint implementation". At a first glance, joint implementation has the great advantage for less developed countries that they receive more developed technologies, especially in energy production. For the advanced developed countries, joint implementation has the great advantage that their balance of CO_2-emissions is improved by means of investments in less developed countries which do not cost as much as they would in an industrialised country. The assumption is rather convincing: The marginal reduction effect on a high level of CO_2-emissions is much larger than on a still low emission level. These advantages, however, are accompanied by considerable disadvantages. *Firstly,* non-optimal technologies are encouraged because better technologies are more expensive than the less efficient technologies and therefore the latter are preferred. *Secondly,* the pressure to innovate in the industrialised countries is lowered by joint implementation because of the lower technological level in less developed countries. *Thirdly,* life style and mode of production, i.e. the technological paradigm, are not questioned. Nevertheless, joint implementation seems to be a rather flexible instrument in the area of environmental technology transfer.

Of course, this kind of technology transfer is also not apt for increasing industrial competitiveness on world markets. Competitiveness is a result of a complex network of factors influencing the competitive position on global markets. They are the result of market constellations and micro-economic competence, but also of industrial policy by public institutions. In contrast to the predominant neo-liberal discourse, industrial policy is an important measure in order to fulfil the target of international competitiveness on a national or macroregional level (e.g. in the case of the EU). The complexity of industrial policy is indicated in diagram 3:

Graph 3
Subjects, Levels, and Instruments of Industrial Policy

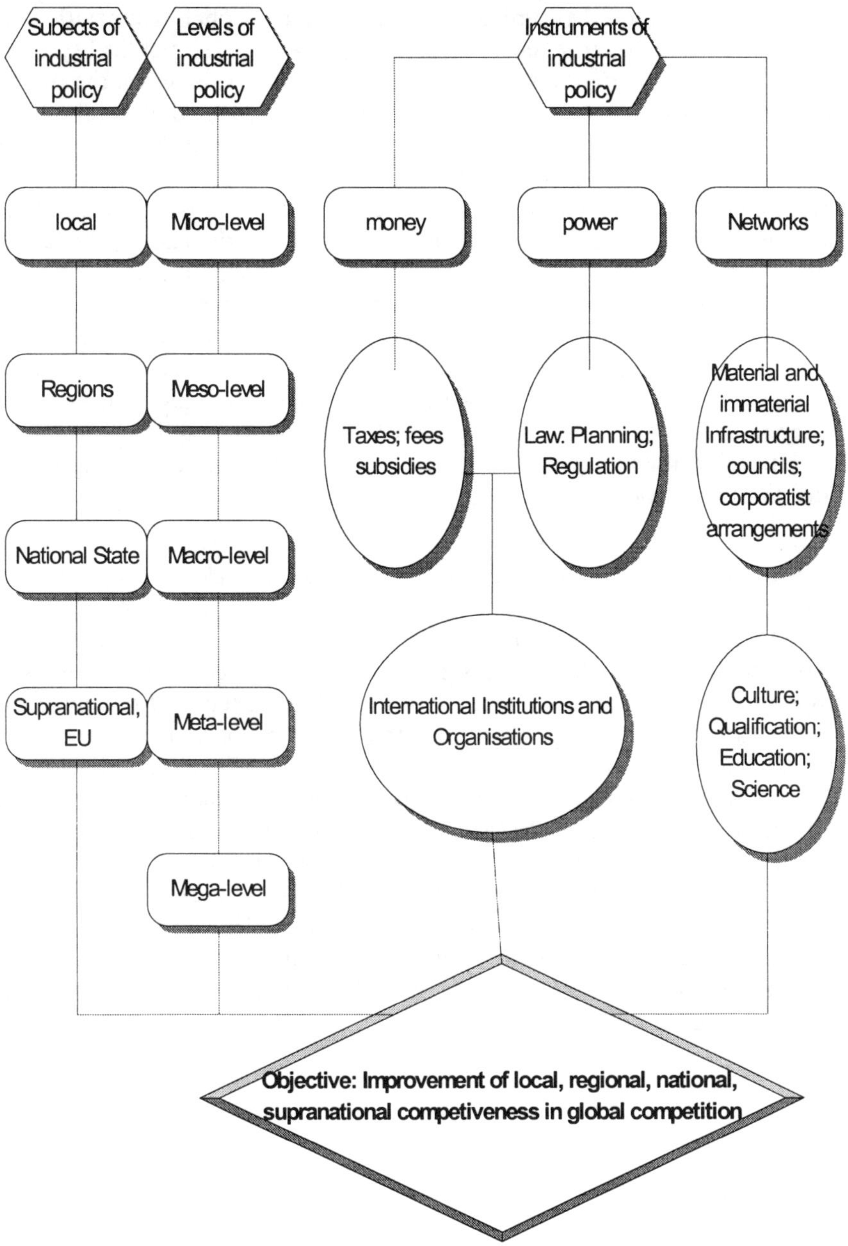

The role of science and technology is only a modest one in the complex grid of those political factors responsible for competitiveness on global markets. These are economic as well as political, social and cultural factors, they emerge from a national society but they are also determined by global factors; they are „in-between“ national culture, economy, society, as well as the political system on one side, and the global markets and regional integration schemes on the other side. It is a decisive experience, especially in Western Europe, that competition has a global reach and that, therefore, global markets are defining the horizon of economic decision makers. Competitiveness, however, can only be created on a local (microregional), national or macroregional level. In Western Europe, politics of competitiveness, therefore, are understood as a priority task on the level of local and regional governments, whereas the national government is responsible for macropolitics, i.e. the regulation of money in the currency competition, the labour market and in industrial politics. But these policies - this is one of the lessons of the European integration process - in many cases only make sense on a macroregional, i.e. supra-national level (The Western European example: Airbus industries). Many modern industries and the technologies applied require supra-national co-operation in order to cover costs, but also in order to mobilise „big science“ and adequate manpower.

The need for scientific and technological co-operation, i.e. of technology transfer between technology givers and technology takers, therefore, offers an important argument for much broader political issues. One of them is the improvement of systemic competitiveness. Since this approach requires some understanding about the system in which systemic competitiveness should be improved, it is also an argument for improved and deepened regional integration on the side of technology-takers. But this co-operation must enhance more than just technological issues, since the development of the qualification structure depends on manpower and manpower is not only an economic factor, but rooted in culture and social networks. In order to mobilise these networks, policies of competitiveness cannot only rely on market mechanisms. The economic policy approach therefore must go beyond the dominant market-oriented neo-liberal recipes. This is indicated in the last diagram (graph 4); competitiveness only partly depends on science and technology transfer. But without science and technology transfer, political and economic co-operation among countries (especially among neighbouring countries) cannot work.

Graph 4
Competitiveness

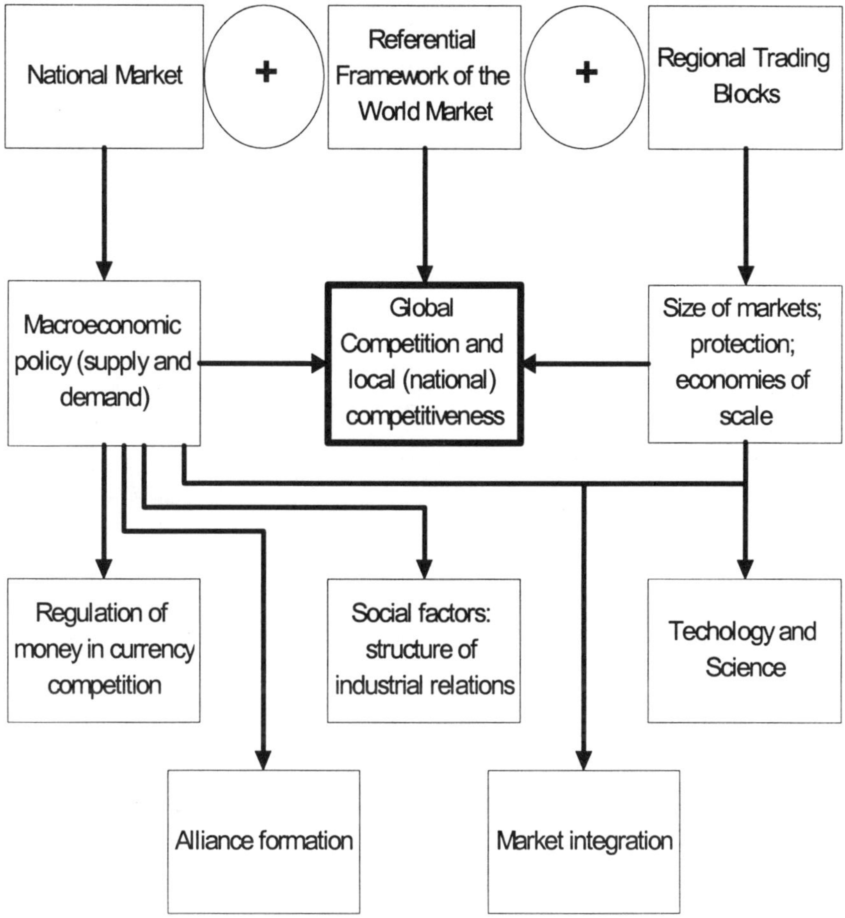

SCIENCE AND TECHNOLOGICAL POLICIES OF THE BALTIC STATES AND INTERNATIONAL CO-OPERATION

MINDAUGAS STANKEVIČIUS
Prime Minister, Republic of Lithuania

Particularly, I am pleased to see here in the auditorium members of governments of the Baltic States and NATO, as well as representatives of entrepreneurship and research. I have a hope that this seminar will offer good opportunities to share the experience in scientific and technological development and international economic co-operation in Baltic and other states.

As I have been told, yesterday many of you had an opportunity to visit some of our enterprises and to find out about their potential and plans for work. Our sincere wish is that this seminar would introduce some dynamics into the current discussion of restructuring and updating possibilities in our economy through expanding international co-operation in the areas of science and technology. Therefore, I suppose, a general outline of the present Lithuanian economic situation could be relevant here.

In the period after restoration of independence in Lithuania the ineffective command economy with planned economic activities has been replaced by market economy practices. The beginning of this decade saw all encompassing restructuring of the national economy and its destabilising effects, i.e. rapid growth of inflation, sharp decline in the production output, the loss of replenishment and repletion functions in the market.

The general market decline was affronted with universal attempts to improve macroeconomicss situation in the country. The main economic policy objectives in the country at present are to stimulate economic activity, to check inflation and support investment processes.

Economic revival became clearly manifest in 1995 first. The net national product in 1994 showed 1 percent growth for the first time since the restoration of independence, and in 1995 it grew by 2.7 percent. This year's net national product continues to grow steadily. By preliminary calculations during the first half of the year it boosted 3.1 percent in comparison with the last year's situation. We now estimate a general rise in the net national product by 4 percent.

The processes of increased economic activities and emergence of market relations are closely linked to the public sector privatisation situation. The growth of economic activity and the formation of market structure is closely related with privatisation processes. The structure of capital in the country has changed critically.

K. Prunskienė and E. Altvater (eds.), East-West Scientific Co-operation, 29–32.

Now the private sector is responsible for up to 65 percent of the net national product. At present, public sector privatisation is carried out only against real money payments.

Rigid monetary policy is maintained in order to curb inflation. Thus, over the last three years, we observe steady decrease of inflation. Last year it reached 35.7 percent which is 33 times less than in 1992. The rate of inflation throughout eight months this year was 10.3 percent which is two times less than in the same period last year.

Monetary policy is reinforced by the Litas Reliability Act. Since April 1994, the Litas is pegged to the US dollar. The Litas exchange rate in other currencies fluctuates in accordance to the US dollars exchange rate differences to other respective currencies. The authorised deposits of the Bank of Lithuania are central for the secure stability of the exchange rate into other currencies. At present, the deposits considerably exceed the required amount to cover the amount of Litas in circulation.

Rapid growth of industrial potential and technological renovation is one of the main concerns in restructuring and revitalising economy and advancing investments. During the last three years, we witnessed a radical shift in the investment source structure. The proportion of national investments shrank, so did the ratio of state and state share- holding companies, while the private capital ratio grew noticeably.

In shaping and implementing capital policies, we strictly abide by the Memorandum of Economic Policy, approved governmental acts, and economic development priorities. In the State Investment Programme, the government has defined areas of urgent priority that command public investments. Among these are energy, transportation and ecology. Several projects in the Health Protection area figure among the priority group, as well. Namely, we speak about the acquisition of medical equipment. The significance of many projects is measured not only against Lithuanian context, but also against the entire Baltic Sea area.

They further integrate into different European projects and programmes. Governmental policy and priorities in the area of investments have been approved by the World Bank experts, international financial institutions, and other foreign experts.

Alongside, Lithuania seeks to attract more foreign investment and use foreign loans with better return. Now, the greater part of foreign capital reaches the country in the form of loans. They come as official support for the Lithuanian government from different international organisations and foreign governments, or in the form of foreign bank loans for different Lithuanian businesses, usually backed by Lithuanian government guarantees. In signing foreign loan contracts, the main requirement is not to exceed the limits fixed in the Lithuanian Economic Policy Memorandum.

Foreign investments are of paramount importance for the further creation and development of joint ventures with authorised foreign capital. Investments of this type reached 400 million US dollars at the beginning of this year. More than 90 percent of all these investments came into Lithuania over the period of the last two years.

More than 90 countries have deposited their money into the Lithuanian economy. Among the biggest investors are Germany, Great Britain, and the USA. European Union States are becoming the biggest economic partners for Lithuania.

No legal acts impose any restrictions for foreign partners income export. Both revenue and dividend export are tax free. According to the Constitution, all regulatory legal acts protect the interests of foreign investors and guarantee that their ownership will not be nationalised, nor expropriated.

In order to attract more foreign and domestic investing, to create more favourable conditions for export of commodities and services both Lithuanian and foreign, and to create a greater number of working places, Lithuania has launched free economic zones. Now we have sound legal basis for the functioning of free economic zones. The Free Economic Zone Founding Basis Act is passed, Siauliai and Klaipeda are legally declared as free economic zones. We have also announced a tender for the draft of a business plan and statute for the Siauliai free economic zone and for the authorisation of the group of founders. We have almost completed working on the Kaunas free economic zone legal documentation and on preparation of the registration projects of scientific-technological parks in Lithuania.

Large investment projects i.e. the "Via Baltica" highway construction, European standard railway network on the Eastern coast of the Baltic sea and high voltage electric lines connecting Lithuanian and Polish energy supply systems are being successfully implemented. Also after developing the Kaunas, Siauliai and Klaipeda free economic zones, the exchange possibilities between East and West would expand greatly. The Baltic States could utilise the industrial potential and geopolitical advantages of their disposition more effectively.

The Lithuanian foreign policy and taxation strategy is aligned with the open economy module for a more rapid integration into the European Union and world trade systems, and with the ultimate aim to attract more foreign investments into the Lithuanian economy. The trade regime of our country is fairly liberal: there are no quantitative margins for import, while import duties are among the lowest in Central and Eastern Europe. Lithuania has already signed free trade declarations with the European Union, EFTA countries and the neighbouring states of Poland, Latvia and Estonia.

At the moment, we are working intensely on the preparation of agreements between Lithuania and CEFTA countries. We have successfully completed negotiation rounds with the Czech Republic, Slovakia, and Slovenia for the preparing of free trade agreements. Next month we expect to start negotiations with Hungary. We hope that the CEFTA agreement will promote economic contacts between our countries.

Structural changes in the area of foreign trade and the steady rise in turnover over the last years, which recently has been growing at about 20 percent, is a striking example of the changed economic activity in our market.

The geography of our foreign trade partners is changing noticeably, now the Lithuanian economy is being oriented towards the Western market, yet a wide spectrum of Eastern partnership is maintained. During the first half of this year, the trade turnover with the European Union, EFTA and CEFTA countries comprised 50 percent of all foreign trade. A relatively large share of Lithuanian CIS trade transactions, i.e. 38 percent, remains because of the energy supplies and import of other raw materials from these countries, mainly Russia. Tendencies, in Lithuanian foreign trade growth, point to the fact that the Lithuanian industrial potential output is growing and production quality is improving.

With the aim to produce export-ready, internationally competitive products, we cannot rely just on political and economic stability of the situation. Today, when the industrial market is being dominated by intellectual work, the investors demand qualified and creative labour, capable of performance in accordance with the newest technological standards.

To occupy a stable place in the European trade market, we have to carry out an active scientific orientation policy that would encourage the appearance of innovative products, advanced technologies, and entire industries built upon this basis. If the first years of reform posed the problem of preserving scientific and technological potential of our country, then now, we need to make the most effective use of it in dealing with economic, social and cultural problems. The Lithuanian scientific potential, namely, twenty nine state research institutes, fifteen higher educational institutions, and a number of technological centres in numerous enterprises, is strong enough to meet the demands of the momentum primarily, through co-operation with national and foreign industry, through active exchange, and through spreading scientific and technological awareness in society.

Thus, for the countries that do not have large natural resources in their possession, science awareness and the education of a technologically minded society is of vital importance. This may become their largest investment into international co-operation.

THE INTEGRATION PROBLEMS OF THE BALTIC STATES: POSSIBILITIES FOR THE FORMATION OF A UNIFIED TECHNOLOGICAL, ECONOMICAL, AND SOCIAL SPACE

BORISAS MELNIKAS
Vilnius Gediminas Technical University

The nature of the problem

At this time, in the Baltic States, there is a complex process of policy, economic, and social transformation - representing the spread of democratic relations, orientation towards the priorities of the universal person's worth, and also the internationalisation of policy, economic, social and cultural development, as well as the formation of working quality level economic market bases.

The transformation processes in the Baltic States are various. One of the basic directions is connected with the Baltic region as a very complex *system* for policy, economic, and social development.

Perspective views toward the Baltic States should be connected with the condition, that the transformation processes in the Baltic States should occur in stages to create conditions for the formation of *a unified technological, economic, and social space in the Baltic States.*

Apart from that, that the Baltic region is *not yet* unified in the technology, economics, and social spheres, there remains the proper conditions for the formation of that type of space. These conditions allow for the Baltic States questions to be viewed as those of *system* development, especially because these questions are remarkably actual in the context of European integration. At this point it is noted, that The integration of the Baltic States into the European Union, the spread of NATO into the East, as well as the co-operation problems of the Baltic States with NATO can constructively be solved only then, if the Baltic region forms a new type of *inter-regional integration and co-operation mechanisms.*

The inter-Baltic State integration cannot be separated from the formation of a unified technology, economic, and social space. It is far more detailed, comprehending the complexity and duality of the formation of such a space, to discuss the array of questions which affect the Baltic region on the basis of its' being a system, as well as on the Baltic States' unified development details.

The following material is devoted to these questions.

1. The Baltic region as a whole: unity and internal conflicts

K. Prunskienė and E. Altvater (eds.), East-West Scientific Co-operation, 33–51.

The conception of the Baltic region as a whole is related to ethnic, social, cultural, political and economic development process *varieties*, as well as the formation of a *set of evaluation criteria* for this region as a whole. The base condition, which must be followed for understanding the Baltic region as a system, is that the Baltic region must be understood as that *whole* together with the characteristic differences and signs of internal conflicts. Also, these signs in the Baltic region are very clear; the analysis of these signs is a crucial link to understanding the Baltic region as an organisation.

<u>The various system indicators of the Baltic region.</u> Understanding in the most general case that the Baltic region - the people residing and their group whole or community - expresses that one real goal and unity, we can conclude that the Baltic region is a system, based on several viewpoints:

- as a whole expressing various internal and external regulation subjects,
- as people and the group whole expressing and implementing various goals and interests,
- as a political, social, economic, or other organisational whole,
- as a formally and informally structured whole,
- as a part of larger geopolitical, economic and other structures,
- as a system having real internal ties, an accumulating social culture, sharing ethnic, historical, political, and economic similarities.

The aforementioned views represent the variety of the Baltic region as a whole, reflecting aspects which show the regions unity, as well as internal conflicts necessary to its development.

<u>Expressions of centripetal and centrifugal force forming in the Baltic region as a system.</u> In the functioning course of every system, there are two types of forces which represent themselves in various forms - centripetal and centrifugal. This team of forces reflects the two system development tendencies, supplementing one another, in relation to that

- the system expresses itself as a unified, constant and inseparable organism,
- the systems separate parts express themselves, to a greater or lesser degree of freedom, in the respect to those rules and limits which conditions the goals, tasks, and interests of the general system as a whole.

The logic behind reflected conceptions of centripetal and centrifugal forces can be used in evaluation of political, social and economic development processes in the Baltic region. Assuming that political, social, and economic development, as well as transformation processes, were conditioned by two rather opposing and supplementing tendencies in the past, are conditioned now and will be conditioned in the future:

- The formation of the Baltic region as a geopolitical, social, cultural and economic community, while the political, social and economic developments rule as much in the Baltic region as in the separate Baltic countries is the Baltic region's *community* and the spread of the *internal integration whole* of the Baltic region,

- The formation of the whole of the Baltic region as separate Baltic countries, while ruled by political, social, and economic development is the *affirmation of every country's internal priorities* and the Baltic region as *averagely integrated* and only surfacely expressing the spread as separate whole communities.

Both of these tendencies are based on a real Baltic region as a geopolitical, social cultural and economic community formation.

In attempt to understand the system and organisational views of the Baltic region as real general development mechanisms and possibilities of further spread, it is necessary to more broadly discuss those actions, which determine the spread of the Baltic country's community as well as of the Baltic region as a whole, and as much those factors which express the opposing regulations to the idea of the Baltic region community and unity.

Arguments determining the idea of the Baltic region community and unity. The basic reasons in favour of the idea of the Baltic region as a *geopolitical, social and economic community* are the following:

1.The shared *geography* of the Baltic countries, signifying the similarities and unity of the Baltic countries in the geographic respect;

2.The similarities of *historic culture* of the Baltic countries, expressing that

- The historical development of the Baltic countries has accumulated many similar factors and facts, indicating analogous reasons as well as clear connections in time,

- The societal development of the Baltic countries is based on mentality, tradition and living style similarities,

- the same historic cultural environment was based on very similar *historical* events, indicating a common cultural history of the Baltic countries; within the affects of the former there is especially those which were related to this region's development in the Russian Empire until 1918, as well as those related to this region's development in the USSR from 1940 until 1990 (the Baltic region's development within the USSR is related to a deep and multifaceted political, social cultural and economic sovietization of life, expressing unified indicators and standards, and in great part determining the similarities of the Baltic countries at this point in time);

3. The shared *political* development of the Baltic countries, expressing that

- The Baltic countries unified and continue to unify along as much real as imaginary *external* dangers; these dangers were and remain to be related to potential political, economic and cultural environment invasions and aggressions in view of the Baltic countries (the Baltic community has experienced many painful historical lessons in relation to negative views towards any aggression or invasion in respect to them: this experience expresses itself as a unifying factor of the whole Baltic region, and has special political meaning); external danger is one of the basic Baltic countries political, and military defence co-operation stimuli,

- The Baltic countries at this time have similar political goals relating to relations with the larger and more influential world countries and international organisations; these political goals include, first of all, guarantees in the areas of security and stability, and they are the Baltic region's motive for political unity,

- the larger and more influential of international political subjects view towards the Baltic region is essentially understood as a unified international political, economic and military defence object (all, without exception, of the larger and more influential world countries and organisations see the Baltic region as being a similar part of Europe - internal politics, economics, and social culture - without paying heed to the more clear and essential specific characteristics, which would be necessary for separate countries); the Baltic region could express itself as a unified political, economic, and military defence object, from the point of view of the larger and more influential international political subjects;

4. The shared *economic* development of the Baltic countries indicates that

- The Baltic countries show a very similar natural, technological, legal, social cultural, political, and other factors of their economic development: in view of the world economic structure, within the whole Baltic region exists a very similar economic structure, and all the Baltic countries need, essentially, the same level of economic development (especially similarities in work load and quality, as well as in work culture), and it is characteristic of all the Baltic countries, essentially, the same economic reforms as well as economic development problems,

- any large modern economic or technological program can be successfully realised taking only the *whole* Baltic region: any one of the Baltic countries is too small for the implementation of a large project, and their markets are too small, therefore, realistic economic or technological advancement in various economic spheres must take place in *all* of the Baltic countries simultaneously (at this point, it is noted that attempt to localise large economic or technological projects in the separate countries at the same time would also cause its ineffectiveness: only the implementation of a complete set project can be comprehended as a modern economic criteria viewpoint towards an effective result),

- The Baltic region is valued as a unified international economic situation;

5. The common *social* background of the Baltic countries indicates that

- The Baltic countries communities' social structures, cultural levels and ethnicity show great similarities,

- Social problems are characteristic of the Baltic countries, relating to the spreading social differentiation and polarisation in society, and also problems relating to the fact that the Baltic countries' social structure and cultural levels are inadequate for a forming economic structure and the extent of economic decreases (the Baltic countries, in their social structural respect, are not ready for drastic economic and technological changes, especially the economic decreases over the past few years);

6. The shared *problematic* tendencies of the Baltic countries and their development in the view of European integration and international co-operation indicates that the Baltic countries raise problems relating to tendencies of the new

European integration and international co-operation: these problems are *general* to the whole Baltic region, including those in the areas of economic and social development and require a view which takes the Baltic region as a *whole*: the aforementioned problems are several:

- the problem of political, economic and humanitarian *support* in the Baltic region is related to the fact that there must be that real and effective *initial* support for European integration and international co-operation conditions, upon which would create opportunities for further spread into international integration orientation processes,

- the international *security* problems, since the Baltic region is at this point considered, in the local respect, but as well as in the real international insecurity and unstable factors: The Baltic region, at this point, expresses itself as a direct source of ecological, criminological, and other hazards and risks to the Western countries (risk of environmental pollution, various types of illegal immigrants, transport of contraband, narcotics business, weapons sales, various other organised crime activities, prostitution, etc.). This is used as a pretext for escalating political tensions between the larger world countries and defence structures, and as a potential regional ethnic and other conflict source; there are other problems in the Baltic region which can present themselves as a threat to international security, as well,

- the spread of the *sphere of influence* of the larger world countries and political and economic organisations or other change problems directly affecting the Baltic region as a whole (Baltic integration into the EU, NATO spread towards the East and other similar problems, which cannot be one way or another, solve without considering the Baltic region as a whole).

The aforementioned, as well as some other reasons reflect the most important arguments allowing that the Baltic region be regarded as a clear geopolitical, social cultural, and economic community. It is also noted that opposing arguments cannot be ignored, indicating arguments that oppose the possibilities of the Baltic region being a community and a whole.

<u>Opposing arguments to the idea of the Baltic region being a community.</u>

The major arguments *opposing* the Baltic region community and whole are:

1. The *geographic* variety of the Baltic countries, indicating the formation of separate geographic conditions affecting geopolitical, economic, as well as strategic military situations; the geographic variety of the separate Baltic countries is not as clear as the similarities, however, the existing differences can objectively be used to oppose the community of the Baltic region and the idea of unity.
2. The *historic cultural* differences indicate that the separate Baltic countries has amassed various differing historic cultural traditions, existing in the past and currently within the societies mentalities development:

- separate nationalities rule the different Baltic countries, having separate cultures, languages, and traditions,

- within the Baltic countries, there is an inconsistent variety of *Christian culture* (Catholic, Protestant, Provoslav, and other) development: this indicates variations in Christian mentality within the Baltic region,

- within the separate countries and regions, there have been unequal affects of various *russifications, polonizations, germanizations, scandinavizations*, and of other similar influences, which are the basis of differences in ethnic cultural integration,

- within the separate Baltic countries and regions, there has historically been different economic structures, various business priorities, and there has existed many different viewpoints through various societal levels towards economic, social, and political conditions: this shows a rather clear difference within the Baltic region (these differences, in part, support differences between historic political, economic, and social life traditions),

- the separate Baltic countries and regions have historically formed very different *ethnic* resident structures: various regional residents, past or present, express clear ethnic specifications, which due to various aspects, indicate different lifestyles and traditions, appearing in the political, social, and economic development in the separate Baltic countries and regions,

- very different national community territorial dispersion was and still is characteristic of the separate Baltic countries and regions: the dispersion throughout the Baltic region territories was and remains unequal, as much in the quantitative respect as in the social levelling respect; the ethnic resident structure, as well as the differences in national community territorial dispersion support rather clear historic cultural distinctions between the Baltic countries,

- the separate Baltic countries and regions differ rather drastically in their community *social* structures, and this supports long-term residential formation of traditions.

3. The *political* differences in the Baltic countries indicate that

- specific political features are characteristic of every country, proving that the difference in political goals and regulations is not consistent with the political orientation of the Baltic region community as a whole,

- the orientation of the larger and more influential of the world's countries toward the Baltic countries may differ little: these differences could by due to many reasons relating to the variety of their interests in the Baltic countries: the existence of those differences also indicates the differences in politics between the Baltic countries in general,

- the Baltic countries, as a part of their region and as international political and economic subjects, are in actuality *competitors*: the political competition between the Baltic countries shows that the region does not display itself as being a unified system.

4. The *economic* peculiarities of the Baltic countries indicating that they have, in the economic respect, in the view of one another, clear *competitive* positions, which were created by the following surroundings:

- the Baltic countries continue to express similar economics and economic structures: it is necessary for the separate nation economies that their branch structures, production force development level, labour productivity and quality, specialisation, as well as created product assortments be very similar; this similarity greatly continues to limit the domestic economic addition and co-operation possibilities and at the same time creates conditions for the appearance of stiff competition as much within the Baltic region, as abroad,

-the Baltic countries do not have a tradition of economic co-operation: as much up until soviet times (until 1940) as under the USSR conditions, the internal economic co-operation in the Baltic region was very little (during USSR times, every Baltic republic had many economic ties with various USSR regions, including those beyond the Baltic region, however those economic ties were formed and kept working exceptionally under the USSR centralised planing and controlling mechanisms); the lack of experience and traditions for creating and keeping internal economic ties within the Baltic region expresses itself as one of the preconditions for the current economic competition between the Baltic countries,

-the very clear current competition between the Baltic countries expresses that each nation strives to represent the economic requirements of the whole region and to attract potential investments for those projects and programs, which essentially would conceal the necessity in use of the Baltic region as a whole: in this respect, it is noted that the Baltic countries compete amongst themselves as investment, as well as economic support attractors for the Baltic region,

- the real working economic development stimulating mechanisms in each of the Baltic countries, are largely geared towards co-operation with Western countries, and not towards co-operation within the Baltic region: this greatly limits the partnership and co-operation of the Baltic countries, at the same time, strengthening interregional competition; this competition, being that the interregional partnership and co-operation are not developed, negatively works against the possibilities, of the Baltic region as a whole, to successfully compete in world markets;

5. The *Social* peculiarities of the Baltic countries ,indicating that each country's communities display a rather clear *closedness* because

- the former social cultural integration connections into Eastern Europe and especially within the USSR, which were supported and working during the Soviet time-period, broke very suddenly; the break of these connections negatively affected the ties between the separate Baltic communities; this situation greatly influenced the closedness of each Baltic country's social cultural community, especially since the former ties between Eastern Europe and the USSR was not compensated with adequate ties to the West,

- the Baltic communities, until 1990, were practically completely separate from the Western social cultural life (only separate individuals and a handful of organisations and establishments had social cultural contact with the West, making a wider integration into the Western social experience and cultural system virtually impossible); during the past few years, the social and cultural rules of the West have been spreading over the Baltics, however, real integration into the Western system is still small scale: this condition very really influences the Baltic communities' closedness, apparent even at present,

- Western policy in the view of the Baltic and other Eastern and Central European countries, was and is oriented towards the fact that these countries' cultural integration into Western systems would occur with limits and would be strictly controlled by the West; the West's rules for social and cultural integration into European and world structures is more political, than practical in nature, and in fact, are only realised in the Eastern and Central European countries' elite level of society (and then only on a limited scale!): this sort of Western policy is due to the fact that in the Baltics (as in other East and Central European countries) will maintain a rather marked community closedness,

- the community closedness in each of the Baltic countries is due to the current perspective of each country, since, first of all, the goal is to integrate not so much into an internal Baltic community (by the by, the Baltic region as a unified and inseparable social organism does not exist), as into the Western community in general; the do or die goal to integrate into a rather abstract and superficially understandable Western community (it is interesting that a Western unified and inseparable community also does not exist) not only greatly slows the real Baltic integration in the social and cultural respect, but also stops the Baltic harmonised community development and unique potential to spread;

6. The *problematic* peculiarities in respect to European integration and international co-operation, indicate that in each Baltic country, there exists a priority of very clear directly relating problems, and general Baltic region or Baltic community problems are not always considered priority.

One of the characteristics of the Baltic region, influencing the integration process problems and complexity of the region as a whole, is that none of the Baltic countries has the abilities and basis to be the *leader* of the integration process, and more so of the Baltic region:

- in the *quantity* respect, (territory, number of residents, natural resources, amassed capital, national products, military and economic power, scientific technical potential, and other parameters evaluating quantity) all the Baltic countries are more or less similar and are seen as weak countries, with small economic and productive power,

- in the *quality* respect, (quality bases representing the community, quality of life, economic level, policy regulations, transformation and reform effectiveness, world image, and other quality indicators) all the Baltic countries are also seen as very similar and not having any basis to lead in respect to the others.

The lack of having a country leader in the Baltic region is due to the fact that the integration process creates a lack of *internal nucleic attraction*, which would be a basis for mobilising and directionally orienting integration process efforts.

The aforementioned, as well as other, reasons reflect the most important arguments, allowing the assertion that the Baltic region integration can really be implemented along limited average models, whose basis in geopolitical, social cultural, and economic community formation will express many developmental tendencies which will oppose the logic of the region's unity. The implementation of these models, on the other hand, could be the basis to rationally balance the views toward the Baltic region as a system full of unity, as much as a whole made up of separate yet similar countries.

2. The Baltic region as a system: the formation of regulating and self-regulating mechanisms

The understanding of the Baltic region's development problems is permanently tied to the formation and implementation of those *regulating and self-regulating mechanisms*, which are designated to guarantee the effectiveness of the region's complex geopolitical, social cultural, and economic system spread. The formation and implementation of these mechanisms require the naming and evaluation of the most important regulating *subjects*, whose interests, in respect to the Baltic region as a system, raising *goals* could be the content for forming a regulating process and regulating mechanisms. This means that the essential question is the comprehension and naming of the regulating subjects and their sought after goals and interests, as well as the most significant regulating subject priority goals and interest evaluation.

The variety of regulating subjects and their goals and interests in the Baltic region.

Any geopolitical, social cultural and economic system development goals, as well as their primary interests, one way or another represent those *real regulating subjects*, which form and realize the *real development strategy*. In this sense, it is very important to comprehend the whole of those economic, political, as well as other structure, organizations, and institutions (formal and informal), which actually fulfill, or have the potential to fulfill, the functions of the *Baltic region's development regulating subjects*.

The real and potential regulating subjects in the Baltic region are various, and are classified according to many criteria.

The classification of the regulating subjects best reflects the variety - classifications upon whose basis it would be possible to describe the subjects, considering their goals and interests and various characteristics. For this reason, it would be best to single out the most important regulating subject goals and interest according to *indicator groups* and classify the subjects on the basis of goals and interests.

The most important goal and interest indicator groups of the subjects are considered to be the following:

1. Indicators, representing the *spatial* characteristics of regulating subject goals and interests:

- goals and interests forming within and affecting the region,
- goals and interests affecting the region which form outside of the region,
- goals and interests forming in and affecting different parts within the region;

2. Indicators, representing the various *natures* of goals and interests:
- political goals and interests,
- military strategy goals and interests,
- economic goals and interests,
- social cultural and other goals and interests;

3. Indicators, representing the *global* expression of goals and interests:
- goals and interests of various global levels and dimensions,
- local goals and interests of various levels and dimensions (the local goals and interests can be chosen on the basis of various criteria);

4. Indicators, representing the expression of goals and interests *in time*:
- constant or long-term goals and interests,
- one-time or short-term goals and interests,
- regularly or irregularly expressing goals and interests;

5. Indicators, representing *formal and informal* goals and interests:
- formal and informal goals and interests of various formal structures,
- goals and interests of various informal structures.

Regulating structures can also be classified by other indicators as well.

It is necessary to notice that the naming and evaluating of the region's regulating subjects, including the question of comprehending the primary development goals and interests, is very *complicated*. This is related to various factors, including:

- that the Baltic region expresses various political and economic forces, which rather markedly affect the whole region as much as the separate countries; these forces distinguish the differences and contrary of their goals and interests, therefore, these forces' large scale activity would cause the *instability*, and various *internal conflict risks*, of the Baltic region's geopolitical, social cultural, and economic systems,

- the real regional development regulating subjects essentially represent those true political, economic, and social *minorities*, in most cases, however, they are able to convince the majority portion of society that they are able to represent the society's majority: this creates situations where the factual goals and interests do not coincide at all with the declared goals and interests, and this could strengthen the opposition between various societal groups, leaders, and members; all this increases the danger of social confrontation and polarization,

- that the regulating subjects of the Baltic region which condition the development can actually be , or are, related to various political and economic forces, working at a more *global* scale than just the Baltic region; this means that various realising goals and interests can express themselves as *reflections* of the actual more global goals and interests,

- that among the actual regulating subjects, there can also be various *informal* structures; this circumstance complicates the comprehension of the expressing goals and interests, and the evaluation of the development of the Baltic region as a system,

- that various marked views of the separate Baltic countries exist - ethnic, political, economic and other; this is due to the rather wide belief that various *local* natures can appear in the separate Baltic countries.

The circumstances presented reflect the variety of geopolitical, social cultural, and economic system regulating subjects and their goals and interests; comprehension of the latter supports the complexity of evaluation problems on the whole of regulating subjects, as well as the seriousness of the problem of solving the questions of the Baltic region's political, economic and social development regulation.

The formation and improvement principles of the regulating and self-regulating mechanisms of the Baltic region as a whole. Every geopolitical, social cultural and economic system development's condition is - the effective functioning of regulation and self-regulation mechanisms. This condition is also true in the case of the Baltic region: this means that it is necessary to comprehend the most important appropriate regulating and self-regulating mechanisms formation and improvement *principles*.

The following are considered to be the most important formation and improvement mechanism principles:

1. The *ecological* principle of the Baltic region's development, expressing:

- that the Baltic region must be comprehended as a complete *biosocial ecosystem*, integrating various biological and social subsystems which exist in the region; this comprehension could guarantee the Baltic region as being viewed as a geopolitical, social cultural, and economic community, and this could be a basis for implementing effective regulating and self-regulating mechanisms,

- that the regional biosocial ecosystem development direction and effectiveness is due to a specific unique *community culture*, whose basis is the implementation of the *adaptation* of people, community, and surrounding biological and social environment, expressing various changes in the environment and community: thanks to the community culture, the regulating and self-regulating means are regulations towards biological, social environment, and community protection and preservation, as well as change; the community culture could cause internal *harmony* and *biosocial balance* in the region,

- that the accumulated *biosocial culture*, integrating *unique* biological and social cultures, must be protected and preserved, expressing the unity, continuity, irreplaceability, and internal harmony of the Baltic region's biosocial content,

- that *Baltic* traditions must gain a foothold in the region's social culture, expressing the region's social cultural unity and the shared mentality of the Baltic residents; the Baltic traditions could be the impulse to spread the integration strengthening *priorities*, and this should be a very positive affect upon the region's political, social, and economic development;

2. The *humanitarian* and *democratic* principles of the region's development, expressing:

- that the development regulating and self-regulating mechanisms must be unconditionally geared towards the general human-valuing priorities; the comprehension and implementation of the latter must be the basis to activate regulating functions and to perform self-regulating acts for all political, social and economic development directions,

- that it is necessary to base regulating and self-regulating mechanisms on actual, and not artistic or "demonstrative" democratic statutes: the implementation of democratic statutes must be related to *unity of legal, duty, and responsibility*, including that actions must guarantee *harmony* and *balance* of goals and interests, which represent various groups within the society,

- that regulating and self-regulating mechanisms must be spread by such means that forced means be eliminated from the practice: in this sense, *non-forcible* democratic priority statutes can be mentioned,

- that the region development's humanisation and democratisation must include all political, social cultural and economic development directions;

3. The *innovative and intellectualisation* principles of the region's development express:

- that the regulation and self-regulation mechanisms must be oriented towards all innovative modernisation and improvement of science and technological advances, as well as the newest technological basis,

- that the regulating and self-regulating mechanisms must promote the intellectualisation of the community, implementation of a high qualification level, and the formation of those types of social and economic structures which would by geared towards a culture of high labour and life, work load and quality improvement and overall betterment of living standards: it is necessary to implement regulation and self-regulation means allowing the creation of modern production and service areas, via high innovation, technology, and qualification indicators.

The listed principles (most important) express essential statutes, affecting the formation of the region development's regulating and self-regulating mechanisms and their further spread: it is important to base and effectively target various regulating and self-regulating *methods*, as well as to complete various regulating and self-regulating *functions* during the practical creation , improvement and realisation. From the Baltic region's aspect, the target means of *strategic planning, projection, and programming* has special priority, attempting to solve actual *integration process promoting and spreading* assignments: it is these means which could be the basis to directionally regulate and control the development of the geopolitical, social cultural, and economic systems.

3. The problems of strategic planning and programming of the Baltic region's integration process

Among the development regulating and self-regulating problems within the Baltic region as a system, there are those which are exceptionally related to *strategic planning, programming, and complete projection method* targeting, and also with the planning and implementing of the region's modernisation plans, programs and projects, on whose basis strategic transformation ideas could be realised. By the by, among the latter, there are those which directly affect the *internal* regional integration, as much as the region's integration into European and world structures.

Principle statutes affecting Baltic region integration processes. Formulating the essential region integration processes and their strategic planning, programming and projection affecting statutes, attention is directed towards the facts:

- that the Baltic region as a geopolitical, social cultural, and economic system internal integration and this system's integration into European and world structures' development content, within, there must be *compromising* decisions, balancing the region's *community* as much as each country's *uniqueness and complete autonomy origins* accumulation,
- that integration process development affecting the region must act in *natural* forms, limiting themselves only to *non-forcible* target methods and means,
- that strategic plans, programs, and complete projects must accumulate whole means, with which, after implementation, concrete modernisation and advanced transformation results could be reached in respect to political, economic, social cultural or other views,
- that the internal regional integration could be comprehended as an *essential and necessary stage* for the Baltic countries and region for integration into European and world structures.
- that the internal regional integration could be an effective enough means, designed to reduce negative results, that at this time in the region's development, the primary expressing tendency is one of *cordon transit type development*, reflecting the ideas of the region as being within the "East -West" cordon transit zone system: this idea expresses the view of the region as a cordon for the West, protecting the West from the East, and vice versa, as well as a "bridge" (or corridor) between East and West; the internal regional integration could speed up the autonomous Baltic development and modernisation, and real strengthened independence.

The presented statutes could be evaluated by preparing and implementing *various thematic* strategic plans, programs and projects; in this case it is necessary to understand that strategic decisions which affect the region's political, economic, and social cultural development are especially important.

The problems of the political and military strategy integration processes. The internal regional integration and the region's integration into European and world structures course and results are influenced greatly by *political* (in nature) integration problem solving.

As it is known, there are *formal* integration oriented structures which exist in the Baltic region currently (Baltic Assembly, Baltic Nation Co-operation Government, and other structures of this nature and level). The creation of these structures and the improvement of their further activities is an important action for the *organisation and regulation* of the integration processes. At this point it is noted, that the presence of such formal structures, and their whole range of actions, could in the future become an effective organisational mechanism, could discern the intensive spread of integration processes in various political, economic, and social cultural development directions.

Evaluating the political, and military strategy (in nature), integration problems and possible solutions, attention is given to several rather important regional political and military strategic development situations:

1. The political and military strategic integration should be comprehended parallel to two different alternative criteria views:

- when the region's (as a whole) integration occurs, orienting the content and forms towards those statutes which are the primary at this point in time within the larger world countries and international organisations' policies: in this case, the region's political and military strategic development, and most importantly, the Baltic region's security would be very much *dependent* upon the larger world countries' policies and their decisions, in respect to the Baltics; in this situation the region's security would depend upon essentially only *external* actions;

- when the region's (as a whole) political and military strategic integration occurs and is very *mobile*, orienting towards such statutes which are, first of all, *not essentially dependent* of external actions: in this case the region , in respect to its political and military strategic development, could gear itself so that it would become *autonomous* and *little or less dependent upon the external* political and military strategic situation actions; in this case the Baltics' security would depend more upon their own policies, of course, not completely disregarding the affects of external actions;

2. The Baltic region integration in the political and military strategic respect should be based on gearing towards *strategic decisions* which have long-term perspectives, are acceptable to all the Baltic countries, and are coordinated amongst themselves, affecting those types of questions such as:

- the Baltic region's, as a *modern community* satisfying the European standards, really socially integrating into the European and world structures, *development vision* (this vision should be an ideal considering realistic implementation),

- the region's political and military strategic development *mission* in respect to biosocial conditions concerning the region,

- the region's political and military strategy *goals*, and possible forms of implementation,

- the *diagnosis* relating to the region's biosocial atmosphere in the view of political and military strategy, and possible *scenario* variations, upon whose implementation the security of the region would be guaranteed;

3. Baltic region integration requires not only the formation of a *common* political development, and legal and economic system spread, but also the regulation of military doctrine and defence systems, and (most importantly) the *harmonisation* of all these common regulations, to guarantee that

- on one hand, that every countries' *sovereign* regulations and real goal possibilities for the realisation of their political and military strategic development be guaranteed,

- on the other hand, that a separate country co-operation and partnership level be reached, that the region would function as a unified, internally *harmonised* system, in term of their *common* political and military strategy development regulations;

4. The region's integration requires intensive formation and implementation of various integration-quickening *projects*; among those as examples:

- projects, designed to *unify and harmonise* all the Baltic countries' political, economic, legal, military and other systems and structures' *on the basis of legal norms and methods*, relating them to those of European organisations (EU) and other European standards,

- projects designed to unify, harmonise and integrate various regulating *information support* systems, as well as *technical and technological equipment, service* and other systems: this unification, harmonisation and integration creates preconditions for the guarantee of the mentioned systems' supplementation, and (most importantly) to seek that every system's potential could be used in any extreme situation as a *reserve or additional* potential giving assistance to another Baltic country: these projects are especially important in respect to military, legal, and other systems,

- projects, relating to various means in *international* political and military development spheres: participation in various international peace-keeping actions, international joint organisations (e.g. establishment of a Baltic Batallion) and other similar projects,

- projects, relating to the preparation of specialists and the expansion of their qualifications, experience, scientific studies and other means relating to political, military strategic development and to their development in the near future.

The listed, and other relating situations are valued through the understanding and solving of the region's political and military strategy integration problems.

<u>The problems with the economic integration process.</u> Every country's priority is economic development. Especially the direction of the economic development, its content, and results, the ability to solve actual problems, and the formation and implementation of economic strategies is conditioned not only by the economic situation in the Baltics, but also largely by the political and social situations and their integration possibilities into European and world structures.

Evaluating the economic development prospects in the Baltics, in is noted that the only alternative (without exception) for all the Baltic countries must be *inter-regional economic integration and a regional integration into European and world economic structures and markets*. In this sense, the integration processes are valued as the region's economic development basis.

Many problems raised concerning the region's economic integration are related to the political development of the region: these questions touch upon the formation of political prerequisites necessary for the spread of economic integration. However, next to these types of questions, another very actual set is those which are related to the solution of the economic content and the evaluation of the nature of the economic atmosphere: especially those of priority conditions:

1. The Baltic region's economic integration should be related to the *specialisations* of the separate Baltic countries, their *co-operation and partnership ideas*, indicating that

- the region's and each country's economic spread should be based on *long-term economic strategies and programmes*, accounting for the separate specialisations and economic co-operation and partnership principle content and form; they should orient themselves as much to their own economic development priorities as much as to those of the whole region, taking into account the different specialisation, co-operation and partnership regulations,

- every nation's economic priorities should represent each and every one of that country's economic branches and spheres, with which that country must, is, or can be a *leader* for the whole region and with which there is an amassed potential for further development,

- each country's specialisations could make the *whole region* more powerful in respect to economic structures which would be oriented towards the large markets of many countries and would be powerful in the high quality of product and high leve of production capacity, allowing them to be competitive in international markets, in turn, activating the economies of the Baltics; inter-regional economic internationalisation would allow the dismissal, or lessening, of the primitive internal competition in the smaller markets,

- there should be the creation and implementation of according economic branch and sphere development *promoting* mechanisms, oriented toward internal co-operation and partnership throught the whole region; there hsould also be created mechanisms for a system which would *harmonise* the region, oriented toward the equal stimulus of the region;

2. The economic integration of the Baltic region should first of all be oriented toward the necessary *modernisational* economic and social problem solutioon, among which are noted the following:

- a complete *production sphere* and *production potential modernisation* of 'high' and new technological implementation basis for the whole economic system, orienting production branches toward world market competition levels and large export potential product production displaying high specialisation and co-operation within the region and high quality, etc. This could guarantee good competition chances for Baltic-produced goods; this type of modernisation must be implemented in those branches where they could actually be rated as priority on the Baltic scale: certain promotional mechanisms should be directed toward those branches (tax deductions, state subsidiaries, etc.),

- a complete *energy system modernisation*, oriented toward the fact that the region have a stabile energy maintenance and that energy sources be used more prudently and effectively,

- a complete *modernisation of the environmental protection system*, so that the less negative affects on the environment could be done in the Baltic region (in the pollution sense) and that less risk of danger could come about,

- a complete *job* network *modernisation and spread*, so that unemployment could be gradually lessened and that there be more jobs requiring *high qualifications and intellectual* resources which would reflect prestige and social safety standards,

- *modernisation of scientific studies and professional qualification nurturing* systems, so that there would be a *reserve of humanity* in the region, reflecting high qualifications, and that the potential of specialists be exploited, being able to formulate, implement and propagate *innovations, modern technologies, and modern products* for all economic activity directions;

3. The Baltic region's economic integration should be related to the real *economic policy* formation and implementation, foreseeing that

- the region's economic integration should be oriented toward social economic development priorities: among which are the *branch* priorities (energy, transportation, etc.), *social* orientation priorities, *state and community security and welfare* priorities (crime prevention, and others), *universal goal* priorities, relating to all spheres of economic social development; these priorities must be formulated considering the *strategic* nature regulations, relating to the region's development,

- a *currency* policy should be implemented, promoting production godd *export* and intensifying production, at the same time guaranteeing the separate countries' national currencies' stability, in respect to hard currency the world over: for this, flexible currency models are required, integrating as much national and Western currency relations, as free floating principles,

- a directional *investment* policy should be activated, so that the economic hearth would be a modern industry in the future spreading on the basis of 'high' technology creation: for this there should be flexible Baltic currency models, integrating national and Western currency with free-floating principles,

- a directional *investment* policy should be implemented in the Baltics, so that the hearth of future economics would be modern industry based on spreading "higher" technological creation and implementation; in the industry spread prospect, there should be a basic regional source of income, creating preconditions for the intense solution for social problems, the rise in living standards and to guarantee further reproduction of spread,

- a *tax* policy should be implemented so that those social economic regulations could coincide with the set social economic development priorities: tax mechanisms should be used to promote development in those areas related to the implementation of social and other priorities,

- a *foreign economic ties* policy should be formed and implemented in view of those relevant priorities, among which are: priorities relating to the region's and nations' economic ties; *foreign investments*; *foreign investment ties*; and those relating to economic ties with various foreign countries and their organisations not in the Baltic region,

- a unified *free economic zone and technological park spreading* policy should be formulated and implemented, integrating the current possibilities to spread free economic zones and various specialised technological, innovative and other similar parks: this policy would enable not only the integration of the region's scientific - technological and production potential, but also would greatly increase the attraction for foreign investments and activate the region with modern economic activities,

- an *economic activation required by the spread of the infrastructure* should be formulated and implemented, oriented toward a trustworthy and stabile working bank, insurance, consulting and other institution network spread; such a network should function accordingly with European and world working regulations,

- in each and every Baltic country, there could and should be the implementation of those means representing the country, related to economics policy whose whole would be oriented toward the spread of the region's economic integration.

The mentioned, as well as other conditions reflect the region's economic integration's very broad, complex, and opposing economic development problem panorama. These problems solutions are feasible only if regulations to a real and broad scale integration activity are formed.

Closing Comments

The comprehension of the Baltic region as a whole is based upon the formation and implementation of political, economic, and social development regulations, and the Baltic region as a unified geo-political and social economic system's development strategy preparation and realisation should be understood as the principle problem of the Baltic countries' integration into European and world structures. The solution of these problems, requiring a deep and complete set of arguments, could be the essential basis to modernise the Baltic countries, their economic systems and societal structures.

The comprehension of the Baltic region as a whole includes these important problem areas:

- the commonality of the Baltic countries and unified activity system and the evaluation of the effects of these actions,

- the specific distinguishing features of the countries, due to their shared problems and integration difficulties,

- external and internal effects and environments, determining their integration prospects and limitations,

- the integration mechanisms and conditions for them to surface and spread,

- the mechanisms, and their dynamic possibilities, for integration into European and world structures by the Baltic region as a whole.

The development of the Baltic region as a whole and its effectiveness, is largely based upon how sensible and effective these development regulations are, and also the self-regulation mechanisms. Regulating and self-regulating mechanism targeting allows the programming and planning of development, and to foresee the forms of rational separate country specialisations and their co-operation and ties with other countries. The basis of these mechanisms could be spread even in the inter-regional integration processes of the Baltic countries.

The systematic studies of the Baltic region as a geo-political, economic, and social cultural whole are very prospective and actual in various ways toward the direction of practical and scientific activity.

Sources

Melnikas, B. "Management and Transformation in Eastern European Nations," p. 155, Lithuanian Military Academy, Vilnius, 1995

Melnikas, B. "Business Environment and Transformation Processes: Eastern European Development," p. 320, Technika, Vilnius, 1995

ENVIRONMENTAL TECHNOLOGICAL POLICY IN LITHUANIA AND POSSIBILITIES FOR THE INTRODUCTION OF NEW METHODS

ASTA LAPINSKIENE
Klaipeda University

Good quality of life for people can be reached only after an ecological balance is achieved. Lithuania, being a European nation, is far behind their Western and Northern neighbours in the area of ecological condition. These nations, under democracy, were able to use the world's science and technology achievements to create favourable conditions for a stable ecological balance, and to preserve the ecological safety of their people. Now that Lithuania is independent, their are more possibilities for bettering the ecological situation. There are some difficult situations, because a free economic market, which is helping to reach and economic balance, cannot solve other problems, such as that of ecology, on its own.

It is believed that moving from a central planning economy to a market economy will improve, in time, not only the economic situation, but also the ecological, because the prices of raw materials will rise, thus provoking a saving of raw material. The country has no resources for financing subsidies for natural materials, such as raw materials and water. The ecological situation is stabilising from the decreasing energy subsidies, along the natural process of economic reform. The recession of industry in Lithuania, from 1990-1995, has resulted in 70% less general pollution in the environment. The market economy will, however, only form the basis for such changes if the possibilities of new technology, consuming less raw materials or energy, are present. The right policy for improving the environment would be to introduce a joint effort between administration and industry. By stopping subsidies for energy, industry will be forced to save on fuel, thus decreasing the air pollution. But, if technology is not available for burning cleaner fuel or for neutralising nuclear waste, then the environment will suffer.

In the process of privatisation, the interests of industry and the ecology are confronting. Industry is permitted to pollute the environment when there is a shortage of resources, if they were not allowed, then the industries would shut down. It is necessary to differentiate between the contamination done in the past, for which the government is responsible, and the contamination of the present, for which the new businesses must account for. In this way, foreign investment can be accelerated. The optimum economic effect, as well as decreasing the damage to the environment, is reached after the establishing of technological control of

K. Prunskienė and E. Altvater (eds.), East-West Scientific Co-operation, 53–54.

industry and by managing the economy by joining the mechanisms of economy and industry. The new free economy should be regulated by acts of law.

Working further with old technology in the present market conditions may permanently damage the biology and the landscape of Europe. We will need a policy of open technology, where we can educate specialists, who will be able to initiate saving of resources and introduce control of technological processes in the future. It is necessary to have, as a basis for economy, regulation of the technological integration into the Lithuanian economy and tax system. These should both be prerequisites.

The market economy will end up with problems, for example, the waste will increase. New forms of packaging, as those found in Western Europe, are introduced, but it is neither the consumers' nor the producers' problem to dispose of it. There is no waste technology in Lithuania. Aluminium, paper, Tetrapak, polyethylene, and other waste is being piled up in mountains. The standards in Western Europe are to either reuse these products or to recycle them into less harmful products. Intensive investment projects for waste disposal must be realised immediately, for if not, then we will face an ecological catastrophe. The latest technologies, not necessarily the newest, can be used for this, considering the fact that this branch of industry does not yet exist. Investment into these types of technologies now will allow it to pay off in the future, by preventing national problems from growing into global problems.

Obtaining the possibilities for ecologically clean recycling is important for the environment and the economy of the country, as it continues to change. Environmental policy and the exterior and interior markets should support each other. Through and integrated socially and economically established waste technology, Lithuania will benefit first hand, but in turn, so will the rest of Europe.

STRATEGIC IMPLICATIONS FOR THE INTERACTION BETWEEN INDUSTRIAL RESEARCH AND DEVELOPMENT AND SCIENCE AND TECHNOLOGY POLICY:
The case of the countries in transition

KLAUS-HEINRICH STANDKE
Internationale Akademie Schloss Baruth/Berlin

K. Prunskienė and E. Altvater (eds.), East-West Scientific Co-operation, 55–85.

1.1 The new challenges
The scientific community constitutes one of the most valuable assets of the cultural heritage of Europe. The dramatic decline of the scientific and technological potential in all countries in the Eastern and the Central part of Europe, calls for immediate and drastic remedies. Otherwise, the scientific research system, as one of the more vulnerable segments of society, faces the danger not only of temporary disruption but also of lasting damage.

Science and technology can contribute significantly to the adaptation process of the countries in transition. They can help to make the economy of those countries more productive and more competitive. They can help encourage export growth and cut imports. They can also help improve the environment. In the *White Paper* of the European Commission on 'Growth, Competitiveness and Employment in the European Community' clear evidence was furnished that the Community has been losing competitiveness.(1) Western Europe's overall economic performance has been declining both absolutely and in relation to its major competitors. The objective of the Community activities within research and development is, therefore, partly to improve the competitiveness of European industry by strengthening the scientific and technological bases and partly to increase the quality of life for the citizens of Europe. If the 'health' of the West European Knowledge System which - though in general rather well funded and rather well functioning - is under increasing challenge from other global players and became therefore a matter of concern to the heads of states and of government of the member states of the European Union, the concerns of those responsible for the 'health' of science and technology in Central and Eastern Europe must infinitely greater.

In view of the increasing global competition. the 'Greater Europe' must mobilise all its intellectual resources and technological capabilities.

It is in this context important to note that the mentioned *White Paper* of the European Commission expresses great expectations on the 'Pool' of high-level scientific and technological competence still existing within the countries of Central and Eastern Europe. It formulates the hope that the 'Potential for Innovation' of both parts of the continent could be mutually exploited and would thus be beneficial for all concerned.(2)

1.2 Scope and focus of this review.
There are many conferences, seminars and 'round table' meetings on the '*newly emerging economies*', '*newly emerging democracies*' or '*countries in transition*'. Among the numerous problems which the Central and East European countries have to solve in their endeavour to become, once again, part of the 'global system', the restructuring of their '*innovation systems*' is usually not obtaining the attention which the matter deserves. If there are well-meaning efforts to look into the difficult situation of the 'Knowledge System' of the countries in Central and Eastern Europe, they are usually concentrating on the '*Science dimension*' and rarely, if at all, on the equally important '*technological aspects*' of the innovation

process of the so-called 'countries in transition'(3)) This seminar in Vilnius is, therefore, adding an important chapter in the large book of the new science to be known as „Transitology".(4)

My following observations are attempting to highlight some of the issues elaborated at length during the seven years old and still ongoing debate on the new role which *Science and Technology* and *Research and Development* respectively have to play in order to assist the countries in Central and Eastern Europe in their dual struggle to (a) transform their economies into market economies and (b) to adapt and hopefully integrate their 'Science and Technology system' into the 'Knowledge System' of the OECD countries. The focus of this paper is on the crucial but not always well understood interaction between the governmental Science and Technology Policy (STP) on one side with the industrial Research and Development (R&D) Strategy of the private sector, on the other side. The interplay of both is a perquisite for the creation of a viable internationally competitive technological base of the countries concerned. *Science and Technology Policy* will be the vehicle, the tool, among the other sectoral national policies to reach this target.

One set of issues in this paper will address specifically the predicaments of former socialist countries in their difficult itinerary towards market economies. The main part of this presentation will deal with the relationship of the Science and Technology Policy of governments on one side and on the other side with the R&D policy of the private sector, with innovation policy, valid in any industrialised country, regardless whether they are already member of the European Union or whether they are, like the Baltic countries, preparing their way for EU membership. I hate to use the word *'transition countries'* for the countries in Central and Eastern Europe, because since 1989 it became painfully clear to the traditional Western industrialised countries and in particular for my own country, Germany, that they too as part of the globalisation of their economies, are all of a sudden in a permanent 'transition'. The Annual Report of UNCTAD for 1996 has shown that the 'world market share' of traditional industrialised countries like Germany, the UK and even Japan continued to decline since 1989 whereas the unemployment figures are climbing. In contrast to this, the developing countries in Southeast Asia as well as the giant country China and others are witnessing a steady increase both in their market share as well as in economic growth.

In the final part of this paper, there will be some conclusions as well as some recommendations for future action.

2.The predicament of the Central and East European countries in general and of the Baltic States in particular.

2.1 The legacy of the past

In the Proceedings of another NATO-symposium on 'Science in Eastern Europe' held in September 1988, we can read: *„Scientific development linked to technology is central to Marxist ideology and to its implantation into government. It is therefore not surprising that science has occupied such an important place in the plans and* *aspirations of the COMECON (CMEA) since its foundation in 1949. It appears, however, that despite such continuity little has resulted to benefit the populations at large. Some spectacular scientific and technological results have been achieved in a few areas, for example space in the Soviet Union - COMECON's dominant partner - but all the nations still rely largely on outdated, sometimes primitive, infrastructures and procedures...*"(5)

It is one of the most unexpected and sobering discoveries in the former COMECON countries that this region which possessed only a few years ago more than 2 million scientists and engineers performing R&D - roughly equal to all OECD countries combined, including the United States and Japan - is today only of marginal interest in the field of science and technology to Western countries. We see here the unfortunate consequences resulting from the separation of R&D facilities (in the former COMECON countries) from the production process, or, how state support was handed out on the basis of institutional reputation instead of on the basis of the technical merits of the specific projects in question. Another related problem was the inflexible Socialist system of central planning, i.e. the so-called *'Comprehensive program...*" of COMECON. Research, Development and Innovation are basically an exploration of the unknown. Many creative discoveries or technical solutions are surprises. Thus, the choice of important research themes at the 'All-Union' ministry level in Moscow or through the "*science and technology five years integrated programmes'* of COMECON worked to the disadvantage of creative, path-breaking research. (6) The so-called *'Basic principles of the international socialist division of labour'* as well as *'The Organisational, Methodological, Economic and Legal Principles Governing the Scientific and Technical Co-operation of the CMEA Member-Countries and the Relevant Activities of the CMEA Bodies'* (7) introduced by COMECON have resulted in production and research facilities which are oversized for the former Soviet Republics once they became again independent.

2.2 The assessment of the situation by Western observers

Many efforts have been undertaken during the last years to evaluate the Science and Technology systems of the former COMECON countries.(8) Unfortunately, no systematic approach to this effect was made and no internationally comparable methodology was used. Therefore, the results are inevitably sketchy. The implementation of the - non-binding - recommendations is for obvious reasons very difficult since they had not only to overcome the vested interests of those immediately concerned, the situation was furthermore made more complicated since most of the evaluated countries had to witness since 1989 not only *one* drastic change of government, but in some cases *several* new teams of ministers and of parliaments with fast changing majorities. As often seen in similar situations, each successor usually rejected the decisions taken by its predecessor. As a result, the drastic shrinkage of funds and personnel in the Science and Technology systems of all former socialist countries is 'not the result of active planning of national priorities and for the corresponding resources needed to accomplish them. Rather it is the consequence of a collection of individual decisions that have left institutes with those employees not necessarily best suited for current and future needs.'(9) The only case, in which the findings of Western evaluation teams have been fully applied is East Germany. As an immediate consequence (see below), the local S & T potential has been drastically reduced (almost by 80%) and most teaching positions have been filled by West German faculty staff.

2.2.1 Industrial Research and Technology

After the fall of communism in 1989, the world scientific community devised bilateral as well as multilateral assistance programs for their colleagues in the former socialist countries. They also offered expert advice for independent international evaluation of the Science and Technology systems in these countries. Western Industrialists also were ready to make special efforts to make the industry in Central and Eastern Europe an integral part of their world-wide Cupertino and marketing efforts. In addition, many industries from Western Europe, from the United States and from other parts in the world regarded the sudden new political situation in these countries as a chance to buy high level technology at cut rate prices. To quote one example, *Siemens* together with *Daimler-Benz* have entered with Russian partners into a Joint-venture with offices in Moscow, St. Petersburg, Munich and Stuttgart with the special mission to screen available technologies in Russia for joint exploitation in Western markets.(10) Western industries were furthermore ready to invest heavily in some of the countries in Central and Eastern Europe and to open offices. Today, after several years of activities in this region - according to opinions voiced at two independently organised but related meetings of top representatives from the West European industry -„*the overall results are limited*“(11): Not only did companies find it difficult to locate the high-tech products and processes in Central and Eastern Europe, but when such data was found, it was even more difficult to integrate it in Western plants and production systems.

Things are not facilitated by the fact that those in charge of R&D in Central and East European countries find it difficult to accurately define what it is that they can contribute to each potential co-operative arrangement.

Moreover, according to the same reports, as far as Co-operation in joint research projects is concerned, the experiences are, in general, not very encouraging. The reasons for this are complex. As an alternative to in joint projects, industry now takes a different approach. Although some co-operation (in joint research and development projects) still exists, most Western industries export their own technologies and know-how, and train East Europeans to work with this. In this context, taking over industries in the East mostly means buying a site at a low price, then replacing it by a completely new factory which functions with modern Western technologies. In other words, the Western industries found that it is cheaper to build new factories than to modernise the old East European ones, which are generally considered to be „*big and inefficient*“.

In more detail, the following findings were reported by representatives from large Western technology-based enterprises:
„In Eastern Europe's R&D one may expect top science, exceptionally good basis education, surprising enthusiasm, ingenuity, especially to overcome problems, eagerness to succeed and willingness to learn and to adapt with respect to language, culture, modern technology. By Western standards salaries are extremely low, which means that carrying out specific tasks can be very inexpensive.

There are, however, substantial obstacles: low technological level, social instability and mutual mistrust of the researchers, poor communications, great difficulties due to differences in culture and language (in spite of the above mentioned eagerness to overcome them) and lack of security with respect to intellectual property rights.

The substantial part of R&D in the Communist countries has been carried out in research institutes rather than in universities and industrial companies, and had been devoted to military objectives. There is generally no notion of Western concepts such as marketing.

At present, due to the economic situation, R&D links to former Communist countries cannot be effectively used as a key to the potentially enormous market. They are, instead, an investment for the future, which seems to be necessary in order to preserve stability and to avoid an even greater brain drain from the East than that has been experienced so far.

Therefore, support by official bodies, e.g. the EC is important. Present EC activities are - possibly due to old established links - directed towards university relations, and thereby lack effectiveness in the improvement of technological relations. Better results could be achieved by:

- *supporting modernisation of the infrastructure in Central and Eastern Europe (e.g. in the field of telecommunications).*
- *improving the 'boundary conditions' of science and technology in these countries (protection of intellectual property, etc.)*
- *promoting exchanges and co-operation in joint projects involving Western industries.“*

Similar conclusions were reached at a Round Table meeting on „*Co-operation with Central* and Eastern Europe“ held in January 1995 by the Industrial R&D Advisory Committee of the European Commission (IRDAC) under the chairmanship of *Rudolf Bertsch* of Siemens. The Committee consists of 21 Industrial Research Directors originating from all 15 EU member states:

„The wealth of new scientific and technical know-how in Central and Eastern Europe is not as 'applicable' as Western industry thought it to be six years ago...
Certainly, opportunities and innovations exist here, but the lack of information, inefficient industrial methods and a focus that is far from any of the Western market priorities, make the bright spots in these countries difficult to identify.

- *Western industries are hampered in their scientific and technological co-operation with organisations from Central and Eastern Europe by the fact that, in particular, the legal situation (protection of industrial property rights) remains uncertain and that there is, in general a poor infrastructure.*

- *The countries of Central and Eastern Europe possess a pool of highly qualified researchers, who are able to adopt quickly to modern technologies. Unfortunately, however, they are primarily specialised in the fundamental sciences (mathematics, physics). The problem is to translate this fundamental know-how into products, and to exploit it to promote industrial competitiveness.*
- *Scientific and technological co-operation with Central and Eastern Europe can only be successful if there is a long-term commitment from the Western side, accompanied by high investments..*
- *Given the previous points, it is very unlikely that the involvement of industry in the EU programme on strengthening scientific and technological co-operation with Central and Eastern Europe will grow considerably."*

The IRDAC Round Table in Brussels, held in January 1995, as well as the EIRMA Annual Conference 1995, attended by R & D Directors from more than 120 industrial corporations from Europe, the US, Japan, India and Korea gathering for the first time in the 30 years history of the association in a country in Central Europe, i.e. in May 1995 in Prague, when debating the „*Global R & D Resources*"(12) reached the same conclusions: Sophisticated technology and the underlying sophisticated R & D is difficult to find in Central and Eastern Europe. Co-operation with those countries means heavy investments and a long-term commitment. For most industries (certainly for the smaller ones), this is outside their strategy since they have a short-term profit oriented view. For these reasons it is felt that in the countries of Central and Eastern Europe, profit from scientific and technological co-operation is still a very uncertain factor.
In sharp contrast to this, the same world-wide operating industrial corporations - regardless of their 'home-base'- are more and more obliged to have at least one, if not several, 'research laboratory antennas' in the United States and increasingly also in Southeast Asia.

2.22 Science

The Western scientific community has attempted to give assistance in various forms to the scientific community in Central and Eastern Europe: Intensified scientist-to-scientist working relations, twinning of institutes, bilateral and multilateral meetings by the Academies of Sciences leading in most of the cases to recommendations for action or real action. Special aid schemes have been devised, i.e. by the European Union in December 1993 INTAS (International Association for the Promotion of Co-operation with Scientists from the Independent States of the former Soviet Union) with an original endowment of more than 20 Mill. ECU. There is also the *International Science Foundation for the Former Soviet Union (ISFFSU)* initiated by Georges Soros with an endowment of 100 Mill. US-$. As a joint initiative launched by the US, the EU, Japan and Canada an *International Science and Technology Centre* was created in Moscow with a starting capital of 75 Mill. US-$. It was meant to provide employment for high-level scientists previously active in the Soviet Nuclear and Military complex in order to prevent their 'Brain Drain' to other countries.

The European Union has enabled the countries of Central and Eastern Europe to use the funds from the largest EU aid programme, i.e. PHARE for almost any science and technology initiative seen as vital by the recipient country. Unfortunately, relatively few projects are funded through this mechanism since other sectors seem to have always higher priorities. In addition to PHARE there are several smaller funds earmarked for science and technology related concerns (TEMPUS, ACE, JEP, PECO, COPERNICUS and others.

The International Council of Scientific Unions (ICSU) in Paris has established a new committee in order to highlight this particular *'problématique'*, i.e. the *Special Committee on Science in the former Soviet Union and in Central and Eastern Europe* which has organised a series of meetings on the complex matter.(14) The last of these meetings, taking place in July 1996 in Paris addressed on of the most crucial issues, i.e. the question of the 'Financing of Basic Research in Central and Eastern Europe and the former Soviet Union' (15). Practically all intergovernmental organisations within the UN system or regionally operating intergovernmental agencies i.e. the Council of Europe, OECD, European Union as well as NATO have put the issue of Science and Technology in Central and Eastern Europe on their agenda. In addition to the international in-depth evaluation exercises of the Science System carried out in particular by the OECD and by UNESCO as well as by the initiatives of some of the Western governments, there all also a series of external evaluations mainly by Western Academies of Sciences. Norway, Sweden and Denmark respectively have undertaken such evaluation individually for each of the three Baltic states (*see chapter 2.4*).

Most recently, i.e. in summer 1996 the US National Science Foundation has undertaken investigation missions to the three Baltic States. Among their findings we find the following observations:

⇒ 'Perhaps most serious from the point of view of science, the Baltic States inherited very large and inefficient Soviet-style industries and institutes that were created to serve primarily the USSR" needs, rather tan local ones.

⇒ Another problem is that the universities are largely devoid of research capabilities, which in the Soviet Union were centred in massive research institutes under the aegis of the Academy of Sciences.

⇒ A still further problem is the ageing research equipment and personnel...One example of this that the average age of scientists is quite high, reportedly 54 in Latvia, as older scientists tend to remain in institutes whereas many younger ones haft sought higher paying opportunities in the private sector or abroad.(13)

2.3 The assessment of the situation by observers from Central and Eastern Europe

This assessment of the situation, seen largely from a West European and from an American point of view, does in essence not differ from the point of view voiced by experts from the countries in Central and Eastern Europe themselves. In the recently published „*World Science Report 1996*“ of UNESCO we read the following evaluation of the current state of science and technology in Central Europe made by a team of Polish experts: „*The system of 'real socialism' gave rise to a very large and generally inefficient research and development structure. This was represented mainly by vast networks of institutes and laboratories managed directly by the sectoral machinery of the centrally planned state economy. It succeeded in receiving a much larger share of the centrally allocated resources than did the system of the academies and higher education.*

The transformation processes of 1989-95 have served to reduce this legacy of 'real socialism': Something that could be seen as a positive phenomenon and as an unavoidable verdict of history. There is, unfortunately, a very dangerous trend in that the liquidation of the old R&D system is also destroying many valuable elements of the old system that were not 'totally rotten'. What is more important is that this process of destruction is not being counterbalanced by the rapid creation of a new R&D system. Instead it is emerging very slowly because of the lack of coherent science and industrial policies designed and implemented since 1989, and a reluctance on the part of the new private entrepreneurship to promote research and development...

A third reason for the low creation of a new system is the reluctance of the trans-national corporations and other foreign businesses to maintain and to develop R&D units in their enterprises in Central Europe.“(16)

Talking about the situation in Russia, in his article entitled 'Russian Science: Snubbed and Sickly', *Sergei Kapitza,* President of Euro-Asian Physical Society observed: „*Fading respect, reduced funding, and the 'brain drain' have brought science in Russia to the vanishing point.“(17)*

2.4 The situation of the Baltic states

In his contribution to the commemoration in 1992 of the 50th anniversary of the formation of the Soviet Union, the former President of the Academy of Sciences of the Republic of the Lithuanian SSR, *J.J. Matślis,* after welcoming the liberation of his country 'from capitalist oppression' in 1940 recalled the state of the economy, culture and science in Lithuania prior to the restoration of Soviet power there' went on to say : „...*The bourgeois-fascist rulers did not care about science, which they regarded as luxury. One fascist Minister of Education even spoke publicly to the effect that science was the privilege of large countries, and small nations should not even dream of it. It was not surprising therefore that the bourgeois-fascist government did not support any real scientific research, and that the funds allocated for laboratory equipment and the purchase of literature and various materials were insufficient even to meet the need of teaching the students. The republic's industry was not interested in science either, for there was little attempt to improve quality or reduce cost prices...“ (18)*

20 years later the Academy of Sciences has been changed into a honorary society of the Lithuanian and foreign scientists, which no longer administrates research. The Academy's vice-president, *Eduardas Vilkas,* presented in 1993 the following assessment of the state of affairs: „...*The Lithuanian system of education and research is not well-positioned to serve national development needs. There are*

several causes for this. Most importantly, the research system, especially that of applied research, was designed to serve the needs of a highly centralised superpower. This, in turn, led to overspecialisation, rigidity, gaps in coverage and a high degree of militarisation. Other factors are well known and common for most former communist countries: insufficient funds, and, as a consequence, a severe shortage of modern equipment; state and party administration of science; restricted access to world science; the separation of research from teaching; weak interest of the economy for technological change. Therefore, an average scientist and average research work were lower in quality than those in the West." (19) According to *Eduardas Vilkas* scientific and technological research, most of which is concentrated in research institutes where the quality is sufficiently high, is in a different situation: *„The Lithuanian State, however, cannot produce reasonable funds for their existence"* Talking about the situation in Estonia. the then President of the Academy of Sciences *Arno Köörna* had this to say: *„In political and industrial spheres there is a strong tendency for supporting R&D which will produce short-term results. The reasons for this are economic difficulties and complicated tasks of transition from the old system to a new one. However, we are struggling to save basic science, which is a relatively high level in the Academy institutes and universities..."(20)* His colleague from the Academy of Sciences in Latvia, *Janis Lielpeters* reached the same conclusions: *„The financial situation of science is made even worse by the fact that neither market nor state administration demands for the results of scientific investigation really exists in Latvia..."(21))*

The Nordic countries Denmark, Norway and Sweden have undertaken in the year 1992 on specific requests of the official representatives of the Science communities in Estonia, Latvia and Lithuania exhaustive evaluations of the Research systems in the three Baltic countries. For example, in the case of Latvia, the Danish Research Councils - based on reports, site visits and interviews - have created 19 expert panels which have presented detailed findings and recommendations in their respective fields of competence. *All the panels have noted a very serious economical crisis in the research system, as in other parts of the Latvian society. They stated that the available funding of research can now cover less than 50 % of what is considered necessary. Furthermore funding from the Latvian Science Council is only given for half a year at a time and covers all expenses incurred in a research project, including salaries, heating, electricity and the whole infrastructure. This prevents planning of the research and provides no incentives for young scientists to remain in the research system. Neither do the low salaries in the universities and research institutes attract the bright young people to enter a scientific career.*

The findings from the in-depth evaluations carried out bilaterally by scientists from the three Nordic countries in the three Baltic countries do not basically differ from each other. The Danish evaluation of the situation in Latvia reached the following main conclusions, which are valid for the two other Baltic States as well:

„Even if the panels understand the very serious economic conditions in Latvia, they wish to point out that unless immediate measures are taken to ensure a stable funding

of the research system, including salaries at a competitive level and funding of the infrastructure, the future existence of research in Latvia is endangered.
The panels are well aware that the solution of these problems may not be possible in the present economy in Latvia. It is therefore necessary to appeal to the international scientific community for assistance in overcoming this serious period of transition.
The panels wish to point out that there is a number of institutes in Latvia, which produce research of excellent or outstanding quality. Many of these institutes are now far too big for the needs of the Latvian society...
The international organisations, as for instance the EC, the UN-system and the Nordic Council of Ministers, could promote Latvian science by providing help in the international marketing of the research capacity and services which are available in Latvia..."(21a)

2.5 The brain drain

The 'brain drain', i.e. the national or international mobility of scientists and engineers is by itself an essential component of scientific and technological development. It has to admitted, however, that in comparison with Western industrialised countries the former socialist countries have been characterised by a certain 'over-capacity' of their scientific manpower. This high *manpower-intensity* aimed to compensate as far as possible for the low *capital-intensity in R&D* (i.e. deficits in scientific equipment and instruments) in the countries of Central and Eastern Europe. Under market economy conditions a reduction of the R&D population was, therefore, inevitable. The issue becomes alarming, when large segments of the scientific population of a given country are deciding or are being forced to emigrate from their country of origin or - this comes almost to the same - if they have to leave their positions and enter unemployment or are obliged to accept positions which have no relation with their scientific skills (External and internal 'brain drain').

The restructuring of the economy in all socialist countries has led after 1989 to a crisis situation in their scientific and technological sectors. It is reported that in all countries the science population has been drastically reduced. Furthermore, the 'image' of a career in R&D seems to have lost its former glamour with the result that qualified young people seek their careers outside the world of Science and Technology.

Whereas in the publicly funded *scientific research* a number of support Programs have been devised by *bilateral governmental* and *non-governmental* initiatives and by intergovernmental organisations (22), the situation of the *industrial R&D system* of the Central and East European countries has been left to itself. Some general statistical data on the 'brain drain' of scientists in formerly government-funded research positions are available, but even less is known on the *exodus* of scientists and engineers formerly employed by industry. In any event, reliable data do not exist on either of the two fields (public research institutes and industrial research institutes).Some figures can demonstrate the dimension of the problem: Poland reports a 'brain drain' because of emigration and change of jobs, especially among young people, of 25 % of its scientific manpower.(23)

The East-West 'brain drain' of portions of the scientific community has to be seen in context with the 'brain gain', i.e. the migration of highly qualified managerial manpower (*'skilled migration'*) into the countries of Central and Eastern Europe.(24))

2.6 Special case study: East Germany

The example of the integration of the former German Democratic Republic, a member state of the former COMECON, as from the 3. October 1990 into the Federal Republic of Germany is certainly not typical for the conditions of the integration of Central and Eastern Europe into the European Union. And yet, there are a number of consequences which should be carefully analysed by the Central and East European countries concerned, when preparing for the full membership in the EU. The East German experience is, rather telling for all other countries of the former COMECON:

In 1989 - according to statistics of the Ministry for Science and Technology of the former GDR - approximately 114.000 Scientists and Engineers were employed in industrial R & D of the GDR. When applying the criteria of the OECD 'Frascati Manual' on the definition of the R&D personnel perhaps the figure of 87.000 would qualify. Regardless of this definition, the fact of the matter is, that in the 'year 7' of the German reunification only 20 % of the industrial R & D staff of the former GDR is still employed. The East German share on the industrial R&D performance of the Federal Republic of Germany is presently appr. 2,4 % The costs for maintaining even at this drastically reduced level the East German R&D enterprise was mainly absorbed by governmental transfers amounting alone in the period of 1991-1995 to 4,5 Bill. DM and for 1996 to 1,26 Bill. DM.

What are the reasons for this dramatic collapse?

1. The Government-controlled Privatisation Agency (Treuhandanstalt) did not regard it as a mandate to introduce some sort of 'structural policy' in order to safeguard jobs or to maintain, even at a reduced level, the 'Knowledge base' of East Germany. Its main task was to form viable companies. Companies which were unable to produce a viable survival plan and could thus not be privatised had to be closed down. Socially acceptable solutions had to be found in these cases for the workforce. Therefore, the Agency did not interfere with market forces not even than when it became obvious that industrial or R&D capacities of long-term strategic importance for the country as a whole were in danger to disappear. The main concern of the 'Treuhandanstalt' has been *„to make the former state-run GDR economy with its vast and cumbersome enterprises competitive in the shortest possible time... This explains,* in the words of the Agency's former President, Ms. Birgit Breuel, *why the Treuhandanstalt operated on the basis of rapid privatisation, thorough restructuring, and, where it is inevitable, closure..."(25))* The German 'Treuhand approach' in privatisation is in sharp contrast to the 'Voucher privatisation' largely used e.g. in Lithuania. But vouchers do neither bring in fresh capital nor new technological know-how,

which is badly needed for the modernisation of the country's industrial stock. It is not surprising, therefore, that President *Algirdas Brazauskas* of Lithuania expressed his dissatisfaction with the results achieved: „*We made a mistake when large industrial units were privatised for investment vouchers...(26)*

2. Most of the approximately 300 industrial 'Kombinats' (VEB's) of the former GDR have by now been privatised or closed. In total appr. 10.000 industrial enterprises of all sizes have been privatised. The capital of those enterprises which have survived is almost exclusively in the hands of West Germans or of Foreigners. East Germans did neither have the managerial skills to cope with the world-wide competition nor were they able to mobilise the necessary capital for the take-over and for the modernisation of the given enterprise. Even if they were in the position to meet somehow these requirements, they would find it very difficult to have access to the equally necessary technological *'know-how'*. In the experience of the German 'Treuhandanstalt' privatisation of former state-owned companies cannot mean 'a big increment in terms of revenue': „*The decisive 'profit' or benefit to be secured thereby is an innovative entrepreneur who is willing to invest in the company and make it competitive on his own account and under his own responsibility"(27)*

The new owners of the drastically reduced industrial operations in the former GDR showed little interest in maintaining industrial R & D units in their East German subsidiaries. They argued that all the needed R&D results could be provided without any extra costs from their central laboratories in the West. (28) To compensate, at least partially, for this almost complete breakdown of the East German R&D system, the Federal Government has created some 15 to 20 Government funded scientific research institutions (Max Planck-Institutes, Institutes of the Fraunhofer Society etc.) which open some employment opportunities for scientists in the so-called 'New Länder'. Other highly qualified East German researchers have found employment in West Germany. But the majority of research staff remains unemployed or was able to find employment outside the field of scientific or industrial research.

As already said, the experiences of the transformation process of the East German economy and foremost the lessons to be learned from the privatisation of its industry cannot be compared to any other of the former COMECON countries. The re-unification of East and West Germany is a political challenge which had to be carried out regardless of the financial costs involved. The *annual* West-East-German transfer amounts to much more than 100 Bill. US-$ (ca. 160 Bill. DM), that is to say one hundred times more than the annual budget of UNDP.

For comparison: The famous 'Marshall-Plan ', i.e. the European Recovery Program (ERP), has disbursed in the years 1948 - 1952 for all West European countries the total amount of 13,9 Bill. US-$. Allowing for inflation, the actual purchase power of the Marshall-Plan-Funds would be less than the annual West-East German transfer.

To give another illustration on the order of magnitude of funds required for the transition from centrally planned economies to market economies: In one of the rare existing estimates of the amount needed for the modernisation of the economy in Central and Eastern Europe, *Jacques Attali,* the first President of the European Bank for Reconstruction and Development (EBRD) has advanced in 1991 the amount of 2.000 Bill. ECU. Roughly a quarter of this figure has already been transferred since 1990 from West Germany to East Germany alone.

Leaving these gigantic figures aside, the East German example allows, nevertheless, to draw some significant conclusions which may be relevant for other Central and East European countries eager to join the European Union:

- The industrial sector which looked competitive as part of COMECON practically collapsed after facing competition from the West. The East German share of the total industrial production of Germany as a whole is presently approximately 5 %. The East German share on the total German exports in 1995 was 9%. The 'balance of trade' between the two parts of Germany shows a deficit of 200 Bill. DM p.a. in favour of West Germany.(29) Incidentally, the same phenomenon can be witnessed in practically all Central and Eastern European countries in their balance of trade with the EU: Whereas in 1989 most of these countries had a small surplus in their trade relations with Western Europe, there is since 1991 an ever widening trade gap in favour of the EU. Germany alone has had in the year 1995 a surplus of 3,5 Bill. DM in its trade relations with countries of Central and Eastern Europe. The East German exports to its former privileged trade partners, i.e. the former COMECON countries went steadily down from nearly 30 Bill. DM in 1990 to less than 5 Bill. DM in 1995.

- The 'knowledge system' of the former GDR (Academy of Sciences and industrial research capacities) has been intensively evaluated by Western experts they have concluded that- with very few exceptions -it is generally in the best of cases only of marginal interest to the West.(30)
- The privatisation process has turned practically all of the emerging new private sector of East Germany into West German or foreign ownership
- East Germany with about 25% of the German population holds appr. 5 % of the private property of the country.(31)
- The official unemployment rate of East Germany is about 16 % as compared to 9 % in West Germany. If all the numerous governmental employment assistance measures would be accounted for, the unemployment rate of East Germany would be close to 25 %.(32)
- East Germany benefits, therefore, from the European Fund for Economic Restructuring (EFRE) to the same degree as the least developed parts of the EU, i.e. Portugal, Spain, Greece, Southern Italy etc. For the years 1994-1999 140 Bill. ECU have been earmarked for the purpose by the EU.(33)

3. Industrial R&D and Science and Technology Policy

3.1 Interaction of Science and Technology Policy with other governmental policies

In the broadest strategic terms, governmental science and technology policy is intended to derive the maximum contributions from *scientific and technical resources* available to the government toward the achievement of a wide range of national goals, including national security, economic growth, public health, energy supply, environmental protection, and quality of life. While economic growth is the principal concern of *industrial research and development,* it is just one of many concerns for *science and technology policy of the government.(34)*

Science and Technology Policy embraces all concerns of the governments for the promotion of the nations technical vitality and competence, policies for allocating resources for research and technical education, and matters of management, organisation and institutional innovation.(35)

The role of government in supporting industrial innovation and in promoting a strong scientific and technological capability has thus become an important tenet of any national policy. Science and Technology policies inevitably intersect with major economic issues such as economic regulation, productivity, and public investment in innovation .

Talking about the present situation of governmental sectoral policies in this part of the world, *Antonin Kuklinski* of Poland had this to say: „*The art and wisdom of long-term policy making is very weakly applied in Central and Eastern Europe. There are no long-term economic policy, social policy, industrial policy, agricultural policy, educational policy, and so on. It is, therefore, not astonishing to find that a really comprehensive long-term Science and Technology Policy (STP) is also non-existent.*

It is extremely difficult, if possible at all, to formulate a comprehensive long term STP as an isolated phenomenon outside the general framework of policy making". (36)

3.2 The role of the government for policy making in Science and Technology

The objectives of the science and technology policy of the government are pursued along several directions, mainly by:

1. Strengthening the scientific and technical infrastructure of the country
2. Initiating programs to pursue specific goals
3. Improving the effectiveness of the scientific and technical system

Science and Technology policies can affect *the infrastructure* e.g.

- by influencing the quantity of graduates in science and technology,
- the balance of effort between basic research, applied research, development and engineering,
- the facilities and equipment to support R&D.

In practical terms, any legislation or government actions that affect these components of the technical infrastructure or 'Knowledge base' of a given country, even unintentionally shape science and technology policy.(37)

3.3 Science Policy *versus* Technology Policy?

The relationships among science, technology, innovation, research and development are fluid. They cannot always be clearly distinguished. What they have in common is, that they are all constituting the *'knowledge system'* or *'innovation system'* of a given country. „Science-based Technology" can be identified as „the major source of economic growth"*(Kuznets, **Modern Economic Growth**, pp.9-10)*

„*Science policy*" is a relatively new phenomenon that emerged during the 1960s and is still in its formative stages. It can encompass both *'policy for science'* and *'science policy'* and *'science for government'(38).* It covers usually the entire spectrum from basic science and research until technological development.

In 1968 it was for the first time acknowledged, during the OECD debates on the *„technological gap"* between de Western Europe and the United States, that the ability to apply scientific discoveries for economic purposes is the most important single factor for the competitiveness of nations.

Consequently 'Science Policy' became in the 1970s 'Science and Technology Policy' and as such it is closely linked to both the industrial policies and the employment policies of nations.

The focus is gradually more on *Technology Policy* than on *Science Policy,* although the line between the two is not always definable. In the OECD countries science and technology policies are today increasingly part of the overall-strategy aimed at increasing competitiveness, hopefully without loosing sight of the equally important aspects 'quality of life' including the environmental issues. One of the key targets of any governmental science and technology policy is, therefore, the creation of the necessary environment for the interaction between science and technology and industry.

3.4 Science and Technology Policy *versus* Research and Development Strategy

What is industrial research and development? In the exhaustive definition of *Herbert I. Fusfeld,* formerly President of the *Industrial Research Institute Inc.* in Washington DC it can be described in terms of the activities performed, and it can be measured in terms of the expenditures and personnel devoted to these activities. The uniqueness, the essence, of industrial research, however, is in its structure. All three aspects - activities, measures, and structure - are needed for the complete picture:

„Industrial research refers to all technical activities conducted within industry that are required to support existing businesses; develop and introduce new products, processes and services; and provide the basis for new businesses. Industrial research includes basic research, development, design, the building of prototypes, and when necessary for the development process, the operation of a pilot plant. It can include activity in those social sciences needed for the successful introduction of a new technology, such as economic analysis, market research, psychological and sociological studies, and even analyses in political science. Industrial research is a convenient and well accepted shorthand that includes the total range of activities within industry known as research and development or R&D". (39).

Whereas academic science is a language *'that knows no frontiers'*, and as such is almost a free good available to any interested party in the world, industrial Research and Development is financed by private or governmental national resources, and its results are used by States and enterprises whose mutual relations are based first and foremost on competition. International funding e.g. from EU or other intergovernmental resources is therefore giving financial support only to research activities which are called 'pre-competitive' or to consortia in which international teams of companies or institutes form task-limited consortia.

What is called for, is a better understanding of the legitimate competing interests of the government, representing the public at large (and thus pursuing long-term goals) and of industry which under the increasing pressures of global market forces has to pursue a R&D strategy aiming at essentially short-term objectives. For obvious reasons industry conducts R&D not to satisfy any intellectual curiosity but only to support its current business and to implement the company's growth strategy.

The required crucial close interaction between the two parts of the equation '*Science*' and '*Technology*' within the national '*Knowledge system*', based on the experience of the OECD countries, can be demonstrated as follows:

„Technological development is only sustainable in the long term with the help of balanced efforts by the public and the private sector. If public R&D and in particular university research, are of a mediocre level, the long-term prospects of OECD economies may be endangered: The environment of firms deteriorates, investors are discouraged and the industrial fabric starts to weaken. But if public efforts are not followed up by private firms, the scientists trained in the universities will emigrate, and the research performed by public laboratories will be stifled or its results will be taken up and developed by firms abroad." (40))

3.5 National concerns in Science and Technology Policy

According to an UNESCO classification (41), the scientific and technological potential of any given country consists in essence of four elements, i.e. the country's:

1. *Human resources,*
2. *Financial resources,*
3. *Information resources,*
4. *Institutional resources.*

Science and Technology Policy is to a large extent conditioned by the *size of the country*:

- Compared to the United States each of the European research systems are considerably smaller. The Baltic countries belong not only to the smallest of the European countries, they are also in the middle of a deep political and economic transformation process. But even without this additional burden, it would be impossible for small and even for most medium-sized and for some of the larger countries to reach a critical size in all fields of scientific and technological endeavour. This forces them, to make choices between different research objectives and to define priorities. On the other hand it makes international co-operations a necessity.

- There are many examples in other small- or medium-sized countries demonstrating that good ideas and originality, based on the creativity of the individual, are not limited to empires or superpowers. The same, of course, can be said about the scientific creativity of small- and medium-sized companies in comparison to large TNC's. Any country that has a modern educational system and an infrastructure for science and technology can be scientifically 'fertile'. A country-by-country comparison of the 'National Science Indicators' using, for example, summary publications and citation counts of scientific papers published in international professional journals reveals to this effect rather surprising results. (42) Still, ideas are not enough. They do have to be transformed into technology , and even many small- and medium-sized countries in the West do not have an adequate R&D infrastructure for translating even some of their good scientific ideas into viable technology.(43) The so-called *'Green Book '* on 'Innovation' published recently by the European Commission reveals tremendous differences in the innovation performance of the EU member states.(44)
- The countries of Central and Eastern Europe will not be able in the foreseeable future to accumulate through their own efforts both the necessary capital and know-how to catch up with the scientific and technological expertise generated by Western countries. As a consequence, the national R&D system and technology policy will have to adjust to a system in which TNC's are accepted by the host country to be major actors. (52) At the same time, all efforts must be made by national governments to make sure that existing endogenous scientific and technological capacities cannot only survive but do receive priority political support for their modernisation.

3.6 International concerns in Science and Technology Policy

When presenting in 1993 its 'White Book' on *„Growth, Competitiveness, Employment - Challenges -The Challenges and Ways forward into the 21st Century"* The Commission of the EU has identified a number of serious deficiencies in the fields of Research and Technological Development. (46)

Among the various measures proposed to deal with this question we find proposals for better co-ordination of the R&D funds spent by member states. It was mentioned that only 13% of all public RTD resources (Research and Technological Development) are directed towards European co-operation in the context of ESA, CERN; EUREKA, the Community Framework programme etc. This means that the remainder, appr. 87% is managed and decided upon independently by the member states of the EU. It is reported that a similar amount being spent mainly by large European enterprises in industrial research and development. A need was felt for the EU member states to have regular exchanges of information on national policies in the field of RTD. Such an exercise might gradually lead to a convergence in the field of RTD policies. It was stressed, however *"that the national RTD policies of Member States are based on the specific needs of their industries and societies. For this reason, there always will remain differences between the national RTD policies of the Member States."(47)*

An open question is, how the countries of Central and Eastern Europe which are already, among other equally important factors, separated by a still widening 'technological gap' from Western Europe, can actively participate in the streamlining process of Science and Technology of the EU countries.

It is for this reason that at an International Conference in Potsdam in 1993, initiated by the Council of Europe, on *'Science and Technology Policy in the Service of the Greater Europe'*(48) it was proposed that in addition to the new science and technology policies of the countries in transition, to elaborate an >*All European Scientific Policy, determining priorities, and in particular including problem areas that cannot be solved by one country in Central and Eastern Europe alone*<. There was no illusion at the meeting as to the difficulties involved in such an undertaking: Since the Member States of the EU for the above mentioned reasons were not in a position to agree on a common unified policy on Science and Technology issues, the equally heterogeneous countries of Central and Eastern Europe would face even larger problems.

3.7 Regional and interregional concerns in Science and Technology Policy

Within the OECD countries - regardless of their size - Science and Technology became during the last decade an increasing strong force in regional development. The regionalisation of research systems is seen in the proliferating initiatives of universities and local or regional governments to develop and exploit the scientific and technological capabilities in their areas for purposes of economic growth, employment and the replacement of mature industries.(49)

On a world scale, regional groupings are also growing. The European Union together with its associated countries in Central and Eastern Europe are the most prominent example. The Central European Free Trade Area (CEFTA), initially launched by the four so-called Visegrad countries, and now open to other countries in the region, is the most recent example.

The question arises whether the participating countries prior to their intended joining of the EU would be ready to co-operate closely on a series of common priority issues, for example on Science and Technology matters. Recent experience shows, however, that the common past in COMECON leading to a politically imposed co-operation among these countries, seems to lead to insurmountable *'mental blocks'*. *Antonin Kuklinski* furnishes an additional explanation. According to him the Visegrad countries find it difficult to overcome 'deep differences in historical trends experienced by the four nations, which have created different national characters and psychologies, and also different feelings of real or fictitious uniqueness.'(50)

As far as the Baltic co-operation is concerned, there seems to be in spite of the obvious advantages particularly on the field of science and technology a certain reluctance. Common strategic defence interests have most recently allowed to reach a common stand on NATO membership. There are also no differences on the common interests concerning a full membership in NATO, however on structural matters like the definition of common utilisation and harmonisation of the science and technology infrastructure, sensitivities seems to exist which are for the time being in the way of a common approach towards the needed restructure of their science system.(51) The former President of the Academy of Sciences of Estonia *Arno Köörna* proposed to this effect a 'common representation' of all three Baltic States in the EC co-ordinating body. He also felt that similar to Finland, Sweden and Norway (the two former countries were at that time not yet EU Member States) bilateral framework agreements relating to scientific and technological co-operation should also be concluded with the Baltic States.(52)

Reverting from the Baltic situation to the larger problem, i.e. intended integration of the two parts of Europe, in which regional development plays an increasingly important role, in this regard, it may be worthwhile to recall the findings of the Potsdam Conference in 1993 organised by the Council of Europe:

The development of harmonious economic and social relationships between the two parts of Europe in so far there are related to science and technology policy depends upon:

a) providing aid to reinforce and extend the scientific and technological infrastructure of the less developed or marginal regions;
b) providing a considerable increase in aid to science and technology programs directly concerns with regional development, whether these are carried out autonomously, in the region itself, or in the context of a collaborative program along existing lines, or in national laboratories;
c) examining the specific characteristics of the various regions of Europe in these respects;
d) evaluating the best long-term balance between policies for science and technology based on existing lines and policies developed from the enhancement of regional differentiation as a strategy.(53)

3.8 The role of the private sector

3.81 Local industry

In Wesern Europe it is reported that in average 58 % of all industrial enterprises undertake in one form or another industrial R&D. Even the very small enterprises (up to 50 employees) indicate that more than half of their innovations are generated from their in-house R&D activities. As larger the company is, their in-house R&D contributes over-proportionally to the roots of their innovations: In large companies (more than 500 employees) 72 % of their new technical knowledge stems from their own R&D efforts.

External sources of information are without any exception of high importance for innovative enterprises. Universities and other external research institutes furnish information to a lesser degree. The larger company, disposing over larger 'knowledge absorption capacities' than small and medium sized companies are able to benefit more from academic research than SME's.

Of high significance is furthermore technical know-how and scientific knowledge obtained from

- the parent companies of subsidiaries
- from the 'joint-venture'-partner
- from supplier companies
- from clients
- from public research institutes
- from technical analysis of other information sources (54)

It is highly recommended that the vast stock of information, statistical data, trend indicators, case studies etc. on R&D policy matters and on innovation strategies on the enterprise level which is available at practically no cost from the OECD and from the EU but also from other intergovernmental agencies as well as from international funding institutions is systematically being screened and adapted to the special requirements of companies in the 'countries of transition'.

The same has to be said about available information and guidance provided for example by the governments of the EU member states, by the US government and by others specially designed for the industry of the countries in question in order to assist them in fostering the innovation process. Since there are not any longer 'national solutions' and 'national recipes' possible, those countries in Central and Eastern Europe can easily benefit from the experiences gained in Western Europe and in other industrialised countries of the world.

Specifically on the enterprise level, companies in Central and Eastern Europe should have all interest to 'plug' into the *knowledge pool* provided by organisations like the *European Industrial Research Management Association* (EIRMA) in Paris, which groups more than 180 research-intensive companies from practically all West European countries . It is deplorable that there is still no member in this organisation from East Germany or from any of the Central and East European countries. EIRMA has been created in 1966 under the auspices of the OECD in Paris as one of the possible answers to close the at that time widely debated *'technological gap'* between Western Europe and the United States. EIRMA has been modelled after the successful and prestigious *Industrial Research Institute* (I.R.I.) with its headquarters in Washington DC. I.R.I. has more than 250 US industrial corporations as members. Similar organisations have been created in Japan, Australia, Canada and in South Korea. These organisations constitute a permanent world-wide non-governmental mechanism for the exchange of experience (and, of course, for personal contacts) among the members of these organisations. This international network on the enterprise level is available to interested industrial companies in the countries of Central and Eastern Europe and can assist them in accelerating their integration into the 'global system'.

Within the framework of the EU there are also numerous industry-oriented advisory committees and associations which produce information of high relevance to companies in Central and Eastern Europe.(55)

Since foreign companies, if they can be attracted at all to invest in Central and Eastern Europe, are reluctant to install in addition to industrial plants costly Research and Development facilities, it is fatal what the former Minister of Science of Poland, *Witold Karczewski,* revealed at a recent International Symposium of the German-American Academic Council Foundation: „*Our private sector is not yet ready to sponsor Research and Development*".(56) *Eduardas Vilkas* reports about the same dilemma in Lithuania: „*...Prospects to get support for applied research from industry are poor, if not hopeless, because of the poor situation of industry itself.*"(57)

Local companies in Central and Eastern Europe - provided they would dispose over the necessary financial facilities - could, of course, instead of generating costly and time-consuming their own technological know-how, procure any foreign technology they wish in order to modernise their production facilities and their production lines. Western companies may be interested in this type of proceedings because of the following reasons (Baranson):

⇒ The decision to release corporate assets, like technology, that have traditionally been considered as of the highest proprietary value is part of a larger and more comprehensive new logistic of both national or trans-national companies as a direct means to earn additional returns.

⇒ The sharing of highly sophisticated, proprietary know-how with a foreign enterprise - either through a joint-venture or through licensing arrangements - is often done to offset the enormity of R&D or production tooling expenditures, or to circumvent non-tariff barriers.

⇒ Western companies may also be interested in what may be called 'measured release of core technology'. This is most common within the process design and engineering industry where the company is interested in selling newly designed technology as extensively as it can and reinvests a portion of profits in developing new generations of technology. In most cases a deliberate attempt is made to retain an essential element of the know-how, without which the purchasing enterprise is unable to develop a more competitive version or to become self-sufficient in the technology.

⇒ Finally, a Western company may be willing sell technological assets that are not considered to be central to the company's business or no longer commercially viable.

In the global trade of technology there is always the danger for the seller that a technically strong and commercially astute foreign purchaser may pre-empt third-country markets, or second-generation R&D by the foreign enterprise may result in the creation of a new aggressive competitor.(58)

But even assuming it would be possible to import the necessary 'Know-how', it cannot be too strongly stressed that without a minimum of company in-house R&D efforts industrial enterprises will not be able to benefit from R&D results produced abroad. It is for this reason that the decline of the scientific and technological capabilities of the Central and East European countries may soon reach proportions which may make them unfit to become equal partners within the global 'knowledge-based' industrial competition. The other option would be, as the case of the transformation of the industrial base in East Germany has demonstrated, that the key industries of the 'countries in transition' will either disappear under strong foreign competition or will be to a large extent under foreign domination. This may be in the interest of job security not necessarily negative. The free movement of capital within the countries of the European Union is, after all, one of the most important political achievements of the treaties of Rome (Chapter 4, Article 67). If carried too far, it will inevitably create tremendous social tensions.

3.82 Trans-national corporations

It has long been recognised that private direct investment through the trans-national corporation is unique in providing from a single source a package of critical industrial inputs: capital, technology, managerial skills and other services necessary for production and distribution. TNC's do not undertake major innovative research without visible prospects of a substantial market - unless they are subsidised. The bulk of R&D financed by the TNC's is done by the parent organisation or in the home country of the parent organisation.(48) Many internationally operating companies see no need for the allocation of funds for international R&D because their domestic R&D programs generate products that can be produced and sold abroad as well as at home.(59)

Jon Sigurdson of Lund in his essay on 'The Trans-National Corporations (TNC's) in former socialist planned economies' brings this strategy to 'the point'.: He says „*the question arises to what extent and in which areas have foreign actors an interest to utilise and expand the domestic R&D structures?* (60). Many TNC's have shifted away from previous strategies of transferring a complete, fully functional business including a R&D facility to a foreign market (61). The necessary 'Internationalisation' of the R&D organisation of TNC's can today be easily achieved without geographic decentralisation. One of the often practised models is called '*centralised internationalisation':* that is,“ the development of systems and procedures to raise the sensitivity of the centralised R&D function to markets, competitors, and sources of technology outside the home base.“(62) In the experience of *Siemens*, globalisation of production has been the driving force for globalising R&D. 12.000 out of the total 47.000 R&D staff of Siemens are occupied outside Germany.(63) The IRDAC Roundtable mentioned earlier confirmed this observation as follows: „*Scientific and technological co-operation with Central and Eastern Europe can only be successful if there is a long-term commitment from the western side, accompanied by high investments.“(64)* Similar is the experience of the *Daimler-Benz group:* 'Foreign investments are a precondition for the sustainable improvement of the economy (including the R&D sector). Such investments, however, need governmentally guaranteed security conditions.“(65) Therefore, *Jon Sigurdson* is probably right when he predicts that the existing R&D structures in Central or East European host countries will only play a marginal role in the newly established subsidiaries of joint ventures. As a more fundamental reasoning, not only valid in the special situation of Central and East European countries, he argues that the acquired enterprise units have initially to be integrated into the global networks of TNC's because R&D is generally conceived to belong to the core competence of TNC's, that should be tightly controlled, while the utilisation of engineering talent can easily be tapped and substituted for high cost engineers in West Europe.(66) Another aspect, equally not applicable alone to Central and Eastern countries, but of importance for any 'host country' dealing with TNC's, has been highlighted by *Alexander Keynan* of Jerusalem: He underlines that International co-operation in industrial technology is generally a complicated matter, especially if co-operation is to be carried out between industries owned by TNC's located in different countries. The government of the host country tries (not always successfully) to assure that it benefits from industrial research done in its country.(67)

4. Conclusions and Recommendations for Future Action

when preparing for the adherence of the countries of Central and Eastern Europe as full members of the EU, the issues concerning the inequality of their science and technological systems with Western standards do not seem to get sufficient attention.

- The performance of basic research in the former COMECON countries is regarded as rather strong. In the past, the absence of market forces and the unknown fierce competition among enterprises the translation of the results of fundamental research into elements of the innovation chain - leading ultimately not only to new scientific knowledge, but also to new technological products - hardly took place. The industrial spin-off from results in military and space research to civil applications, for many years one of the most important sources of technological innovations in Western countries, was practically unknown.
- The severe economic reforms which all Central and East European countries are facing have had already serious consequences for the viability of the R&D systems of these countries. This will inevitably have a negative long-term impact on the economic take-off so badly needed in Central and Eastern Europe. An ICSU workshop on the financing of basic research in Central and Eastern Europe held in Paris in July 1996 and attended by high-level representatives from 16 Central and East European countries reached the conclusion: *„In general, the governments consider basic research <a needless luxury>...Participants noted with concern the general trend of their governments to pressure for the expenditure of limited financial resources on applied research, where results can be more quickly demonstrated, rather than on basic research."(68)*
- Since neither government nor the private sector can mobilise the necessary financial sources to preserve, even at a reduced level, a viable 'Knowledge system' during the period of transition which Eastern Europe is going through, 'awareness raising efforts' efforts should be made by the international community. The governments concerned should realise that their entry

conditions into the European Union will become even more difficult if their most important asset, i.e. the .highly qualified manpower within the national research systems, has been dispersed. The State Committee for Scientific Research of Poland has thus rightly observed: *„The system Science-Technology-Economy should be recognised as the leading element in the development of Central Europe and its integration with the European Union."(69)*

- One of the strongest arguments in the increasingly heated debate on the opening of the European Union for interested countries in Central and Eastern Europe is the expectation that these countries will be belong to the fastest growing markets for technology in the years to come. All of them do have presently a huge demand in modern technology but do not dispose of the necessary purchase power. But unlike many developing nations of the still so-called 'Third World' they dispose of a large potential of qualified scientists and engineers as well of skilled manpower and this is, after all, the most precious global resource.
- What seems to be called for is a series of measures aiming to mobilise the still considerable scientific and technological assets of the countries of Central and Eastern Europe within the framework of the emerging market economic system. Of equal importance is a set of measures seeking to link the new national R&D systems on all levels with the R&D of the Western countries.
- The most important political decision to be taken by governments concerns the funding of the science and technology base of the country. In average, the West European countries are spending more than 2% of their GDP on R&D. Within the EU of the 12 in average 0,92% of the BIP is being spent on R&D (1992) The countries of central and Eastern Europe, operating on a much lower GDP level in absolute terms, should aim at a target which is not much lower than that. The basic research sector should have an identifiable considerable portion of the total R&D funding, i.e. 1.0 to 1.5% of their individual GDP. (70)
- The governments have also to decide on the ratio *Basic Research- Applied Research-Development.* Japan, for instance, spends only 12,6% of its R&D budget on basic research, the US 14,6 and the European countries Germany and France 19,8% and 20,3% respectively.
- Within the expenditures on Industrial R&D there are marked differences between the OECD countries. The 15 countries of the EU have obtained in average 13,5% of their total industrial R&D expenditures in form of government subsidies. The US: 29,3%, France 19,5%, UK 14,8%, Germany 8,4%. These figures demonstrate the interaction of governmental oriented industrial policies with the orientation of the R&D performance of the private sector. In addition to this there is a whole series of governmental instruments to subsidise the manufacturing sector and to influence R&D: Subventions, interest-reduced loans, ***government securities,*** government joint ventures, tax incentives, mixed forms of some or all of them. (71)
- Another important factor of the national 'Knowledge base' is the location of the R&D Scientists: in the United States almost 80% of the R&D staff is employed

by the private sector, in Japan nearly 40% and in the European Union (average) some 25%.

- Integration within the knowledge-transfer system of the OECD countries.
- Screening achievements in the basic sciences and linking them with market-related mechanisms.
- Participation in the traditional multilateral intergovernmental programs of the EU, such as EUREKA or COST, as well as in the newly launched special EU R&D multi-country programmes, such as 'The PHARE R & D Networking" DANTE (Delivery of Advanced Network Technology to Europe), „The PHARE ACE Programme (Action for Co-operation in the field of Economics), R&D Networking: COSINE (Co-operation for Open System International Networking in Europe).

References

1. Europäische Kommission,Weißbuch 'Wachstum, Wettbewerbsfähigkeit, Beschäftigung - Herausforderungen der Gegenwart und Wege ins 21. Jahrhundert, Luxemburg1993, p.63

Europäische Kommission, Grünbuch zur Innovation', Luxemburg 1996, p.5

2.Europäische Kommission, Grünbuch..., op. cit., p. 97

see also: Fiedler, Heinz, Innovation Centres in Central and Eastern Europe, Berlin 1992

3.for example:

⇒ *Trieste Conference on Scientific Co-operation with Eastern Europe, 26.-27.3.1990*

⇒ *UNESCO High-level Colloquium on „Science and Technology for the Future of Europe", Berlin, 25.-27.9.1990*

⇒ *The Role of Academies as Learned Societies in the New Europe, Academy of Sciences, Stockholm, March 1992*

⇒ *UNESCO International Seminar on 'Organisational Structures of Science in Europe', Venice, 27.-29.4.1992*

⇒ *EIRMA Round Table on: Eastern European R&D Resources - a challenge for EIRMA members, Saint-Paul de Vence, 7.-8.1.1993*

⇒ *Council of Europe Conference on 'Scientific Policy in the Service of a Greater Europe', Potsdam, 28.-29.1.1993*

⇒ *ICSU Conference on 'The role of Science in Rebuilding Central and Eastern Europe', Leeds Castle, April 1993*

⇒ *OstWestWirtschaftsAkademie and Bundesverband der Deutschen Industrie, Conference on East-West Research and Technology Co-operation, Berlin, 19.5.1993*

⇒ *KBN-University of Warsaw International Conference on 'Science - Technology - Economy: The Experiences of and Prospects in Central Europe', Warsaw-Konstancin, April 1994*

⇒ *IRDAC Round Table on 'Scientific and Technical Co-operation with Central and Eastern Europe', Brussels, 18.1.1995*

⇒ *EIRMA Conference on 'Globalisation of R&D', Prague, 31.5.-2.6.1995*

⇒ *ICSU Workshop on the Financing of Basic Research in Central and Eastern Europe and the former Soviet Union, Paris, 15.-16.7.1996*

4. KBN

5. Sinclair, Craig (Editor), The Status of Civil Science in Eastern Europe, Dordrecht 1989, p.XIV
see also: Sinclair, Craig (Editor), The Status of Soviet Civil Science, Dordrecht 1987
Darvas, György (Editor), Science and Technology in Eastern urope, Harlow 1988
Standke, Klaus-Heinrich and Richardson, Jacques (Editors), Science and Technology for
the Future of Europe: New Forms of Co-operation between East and West, UNESCO,
Venice 1991

6. Standke, Klaus-Heinrich, Growing East-West Technical Interdependancy, in:
Fusfeld, Herbert I. (Editor), Effective use of Global Technical Resources, Rensselaer,
N.Y.1992, pp.25-39.

7. see e.g. the 'Collected Reports on various activities of bodies of the CMEA (COMECON)' published annually by the COMECON secretariat in Moscow. See also: The
Council of Mutual Economic Assistance: 25 Years, Moscow 1974, p. 62
Kaser, Michael, COMECON, London 1965
Bethkenhagen, Jochen and Machowski, Heinrich, Integration im Rat für gegenseitige
Wirtschaftshilfe, Berlin 1976 .

8. see reports of meetings listed above in # 3

9. Bikalas, Norbert M., Science in the Baltic States, NSF/Europe Report No.82, June 1996, p.2

10. Schmidt-Bischoffshausen, F und T-Kooperation der Daimler-Benz AG in der GUS, in:
Bundesverband der Deutschen Industrie and OstWestWirtschaftsAkademie (Editor),
Berlin 1993

11. Cooperation with Central and Eastern Europe, in: IRDAC Science and Industry News,
No.6, Spring 1995 p.6
EIRMA Representatives Round Table Meeting „Eastern European R&D Resources - A
Challenge for EIRMA Members", Summary of Proceedings, EIRMA/93.4, January 1993

12. EIRMA, Globalisation of R&D, Conference Papers Vol. XLIV, Paris 1995

13. NSF/Europe Reports Nos. 83, 84 and 85 dealing with 'Science in Estonia, Latvia and
Lithuania' Paris, Summer 1996
14. see above # 3
15. Glenday, Christine, Report on Workshop on the Financing of Basic Research in Central and Eastern Europe and the former Soviet Union, ICSU, Paris 15-16.7.1996
16. Kuklinski, Antoni and Kacprzynski, Bogdan, Science in Central Europe, in: UNESCO World Science Report 1996, Paris 1996, pp.81-91
17. Kapitza, Sergei P., Russian Science: Snubbed and Sickly, in: The Bulletin of the Atomic Scientist, May/June 1994, pp.47-52
18. Matulis, J.J., Soviet Lithuania and Its Science, in: Academy of Sciences of the USSR 1922-1972 (Editor), Moscow 1972, p.231
19. Vilkas, Eduardas, Science in Lithuania, in: Standke, Klaus-Heinrich (Editor), Science and Technology in the Service of a Greater Europe, Frankfurt-New York, 1994, pp. 167-168
20. Köörna, Arno, Science in Estonia, op. cit. , pp.139-144
21. Lielpeters, Janis, Science in Latvia, op. cit., pp.163-165
21a The Danish Research Councils, Latvian Research - An International Evaluatian,
Copenhagen December 1992, pp. 3/4
22.Biggin, Susan and Kouzminov, Vladimir (Editors), 'Brain Drain Issues in . .....Europe',UNESCO-ROSTE, Technical Report No.15, Venice 1993
UNESCO-ROSTE, 'Brain Drain Issues in Europe: Cases of Russia and Ukraine, Technical . Report No.18, Venice 1994
23. Kuklinski, Antoni (Editor), Science-Technology-Economy, Warsaw 1994, p.98
24. Hillmann, Felicitas und Rudolph, Hedwig, Jenseits des brain drain - Zur Mobilität westlicher Fach- und Führungskräfte nach Polen, WZB-Bericht Berlin 1996
25. Breuel, Birgit, Mandate and activities of the Treuhandanstalt, in: Privatisation - Together into the Social Market Economy, Berlin 1992, p.166
26.see above # 13
27. Breuel, Birgit, op. cit., p168
Rembser, Josef, Die Integration der wissenschaftlichen Gemeinschaften in West- und Ostdeutschland - Gedanken über das deutsche Beispiel, Bonn o.J.
Rembser, Josef, Wirtschaftliche, technologische und wissenschaftliche Entwicklung in Osteuropa unter besonderer Berücksichtigung von Erfahrungen aus dem deutschen Einigungsprozeß, Manuskript, Wintertagung 1994 der Wirtschaftsgilde e.V., 4.4.1994
28. Hahn, Carl J., Einführung, Industrieforschung in den neuen Bundesländern - eine Zwischenbilanz, wir., Berlin August 1996, p.9

29. Standke, Klaus-Heinrich, Zur Rolle von Wissenschaft und Forschunhg bei der Implementierung der europäischen Strukturförderung, in: Piehl, Ernst (Editor), Europa in Ostdeutschland, Bonn 1996, pp. 295-300
30. Standke, Klaus-Heinrich, Zukunft der Forschungslandschaft Brandenburg , in: Anhörung des Ausschusses für Wissenschaft, Forschung und Kultur des Landes Brandenburg, Schriften des Landtages Brandenburg, Heft 1 (1992) pp.40-49
31. Standke, Klaus-Heinrich, Politische Herausforderungen und weltwirtschaftliche Chancen des europäischen Einigungsprozesses, in: Zeitschrift für Vermressungswesen, No.1, Vol.121, January 1996, pp.3-10
32. Frankfurter Allgemeine Zeitung, 26.9.1996
33. Wulf-Mathies, Monika, Europäische Strukturförderung in Ostdeutschland - Zwischenbilanz und Perspektiven, in: Piehl, Ernst (Editor9, op. cit. , pp. 33-40
34. Fusfeld, Herbert I., Industry's Future - Changing Patterns of Industrial Research, Washington D.C.1994, p.219
35. Branscomb, Lewis M., Science, Technology and Public Policy Research..., in: Inose, Hiroshi et al. (Editors), Science and Technology Policy Research..., Tokyo 1991, p.25
36. Kuklinski, Antoni,
37. Fusfeld, Herbert I., op. cit., p219
38. Branscomb, Lewis M., op. cit., p.26
39. Fusfeld, Herbert I., op. cit., p.4/5
40. OECD, p. 43
41. UNESCO, An Introduction to Policy Analysis in Science and Technology, Science Policy Studies and Documents No.46, Paris 1979, p.
42. Kozlowski, Jan, The Polish Science (based on Science Ciation Index), Warsaw 1994
43. Keynan, Alexander, The United States as a Partner in Scientific and Technological Co-operation: Some Perspectives from Across the Atlantic, Arnegie Commission on Science, Technology, and Government, New York, June 1991, p.31
44. Europäische Kommission, Grünbuch..., op. cit.
45. Sigurdson, Jon, The Trans-National Corporations (TNCs) in former socialist planned economies, in: Kuklinski, Antoni (Editor), Science-Technology-Economy, Warsaw 1994 p.110
46. Europäische Kommission, Grünbuch..., op. cit., p.94

47. IRDAC Science and Industry News, No.7, Autumn 1995, p.6
48. Standke, Klaus-Heinrich (Editor), Science and Technology Policy in the Service of a Greater Europe, op. cit., p. 20
49. OECD, Science and Technology Policy Outlook, Paris 1988, p.25
50. Kuklinski, Antoni, The Transformation of Central Europe, in: Kuklinski, a.(Editor), op. cit., 498
51. Bikalas, Norbert M., op. cit., p.2
52. Köörna, Arno, Science in Estonia, in: Standke, Klaus-Heinrich (Editor), Science and Technology Policy..., op. cit., p143
53. Standke, Klaus-Heinrich (editor), Science and Technology..., op. cit., p.20
54. Europäische Kommission, Grünbuch..., op. cit., p.103
55. e.g. IRDAC and others
56. Karczewski, Witold, Science and the Transition Period in East-Central Europe, in:
DAAK/GAAC (Editor), The Future of Humanities and Social Sciences in Central Eastern . European Countries with Consideration of Experiences from the German Unification
Process, Bonn-Washington 1996, p.50
57. Vilkas, Eduardas, Science in Lithuania, in: Standke, Klaus-Heinrich (Editor) Science and
Technology Policy..., op. cit., p.168
58. Baranson, Jack, Technology and the Multinationals, Lexington/Toronto 1978, p.5-6
United Nations, Mulinational Corporations in World Development, New York 1973
United Nations, Trans-national Corporations in World Development: A Re-examination,
New .York 1978
Reiffers, Jean-Louis et al., Trans-national corporations and endogenous development,
UNESCO Paris 1982

59. The Conference Board, R&D in the Multinational Company, New York 1970, p.54
60. Sigurdson, Jon, op. cit., p.111
61. Stopford, John M., Competing globally for resources, in: Trans-national Corporations,
Vol.4, No.2, August 1995, p.34-57
62. Carnegie-Bosch Institute, Managing International Research and Development, Pittsburgh 1994, p.3
63.Danielmeyer, Hans-Günter, The Global R&D Operations of Companies in the United

States, in: EIRMA Globalisation of R&D, op. cit., pp.21-33

64. IRDAC Newsletter on Science and Technology, Co-operation with Central and Eastern Europe, No.6, Spring 1995, p.6
65. Schmidt-Bischoffshausen, op. cit.
66. Sigurdson, Jon, op. cit., p.111
67. Keynan, Alexander, op. cit., p.6
68. Glenday, Christine, Report on Workshop on the Financing of Basic Research, op. cit.
69. Kuklinski, Antoni, Introduction, Science-Technology-Economy, Warsaw 1994, p.11
70. Glenday, Christine, op. cit.
71. Europäische Kommission, Grünbuch..., op. cit., p.89

ON ECONOMIC CULTURE AND DEVELOPMENT WITH SPECIAL REGARD TO THE ROLE OF SCIENCE AND TECHNOLOGY IN THE PROCESS OF TRANSFORMATION IN CENTRALEASTERN EUROPE

ARNDT HOPFMANN
Free University Berlin

The idea that attitudes matter has been a part of development thinking all along the post-war decades. Obviously, any development strategy is »based on culture«, if only because it is not possible to operate outside culture, as long we understand »culture« as all learned and shared behaviour, ideas, and norms of interaction between people of a certain nation or region. Thus also the western attitude to development (»western developmentalism« - Nederveen Pieterse 1991) is based on specific cultural patterns. »Growth-oriented strategies have been based on the culture of economism. Structural Adjustment Programmes reflect a culture of economic globalism« (Nederveen Pieterse 1995: 184).

This has two implications for the East-European transformation process. Firstly: if we understand this process as re-integration of the post-state-socialist countries into an international economical and political system which is widely dominated by the globalization of western cultural values - of which the »rules of economic behaviour« probably have the greatest significance - then the transformation process will be seriously influenced by western (economic) culture. Secondly: since the countries of Centraleastern Europe have to carry out the transformation under concrete historic constellations, their own cultural heritage will influence the re-integration process.

The following considerations will focus on the role of science and technology in this interaction of different cultural environments during the transformation towards market-dominated economic structures in Centraleastern Europe. Firstly we will try to shed some light on the general characteristics of the capitalist mode of production and the role of science and technology. The second question to be dealt with concerns the problem whether or not there is some justification to differentiate between »western« and »eastern« styles of development in science and technology. Thirdly we will discuss some possible consequences of different traditions in the field of science and technology for future developments.

I

K. Prunskienė and E. Altvater (eds.), East-West Scientific Co-operation, 87–91.

The expansion of capitalism round the globe with its starting point in Western Europe has been perceived as being synonymous with the process of industrialization. This does not mean that all the tools necessary for this success story are originally inventions of a capitalist society. But such indispensable instruments of global expansion like the compass, gun powder or the knowledge of overseas navigation that are originating from different, mostly Asian societies has been introduced into a special scientific environment, where their use was turned away form artistic or religious purposes to tools of production and the extension of trade relations in order to colonize the world. In general, after the period of small-scale enterprises based on craftsmanship - the childhood of modern industries - science and technology became more and more an important factor of capitalist production. This was described by Marx - still the most brilliant analyst of this epoch - as the »transformation of science into a productive force« as the real basis for the development of the Great Industry. »Fact is, that the technology which made possible navigation of the world depended on a science which produced other technologies and other sources of wealth« (Armour 1995: 85).

Thus we have some justification to say that all modern science and technology is of capitalist origin, deeply influenced by Western economic culture and especially by the Western style of thinking. This not only creates path dependency regarding further technological development (see North 1990), but also acts as a dominant factor of attraction to late comers in industrialization. All the development strategies of the so-called peripheries of the capitalist world system aim at catching-up with the scientific and technological success performed by the advanced capitalist nations. But in order to reach this goal, they must try to combine their national or regional economic culture with the achievements of modern western-style science and technology.

II

If we bear in mind what was stated above, we will easily reach the conclusion that all economic culture in different countries - or more appropriately regions - is of mixed and hybrid nature. Essentially, economic culture consists of values and behaviours that are orientated on modern development aims, i.e. industrializiation. And there are also specific regional traditions that determine in which way, or within which social structures industrialization should be carried out. What made the case of Centraleastern European countries exceptional is their recent history of State-Socialism . Generally, we can understand this historic period as the attempt to carry out an alternative way of »catching-up development« avoiding most of the obviously inevitable »social hardships« of capitalism, i.e. unemployment, aggravated social differentiation, poverty etc.

Which are the main specific features of a state-socialist economic culture with in consideration of science and technology?

- first of all we should mention the impressive achievements in space research, that are, of course, misused for strategic military purposes (but this belongs to the Cold War in general). But space research in the former USSR - including the commitments of all other state-socialist countries - was much more »earth-related« and hence »economy-related« than in any other Western nation. Today many Western states try to participate on experiences of the former Soviet Union / Russias in the utilization of permanent space stations.

- secondly, there was a »culture of improvisation« in the economy - a »culture of asymmetric answers« to the challenges of technological progress in the »West« - which was mostly due to the fact that state-socialist countries were not able to draw on the seemingly unlimited resources of the world market. However, this also means that the Centraleast European countries dispose of a more or less developed structure of basic and of production-related research capacities what will soon prove to be very important.
- thirdly, as we know, there was a strong political will to combine the results of scientific and technological progress with social parameters; in the light of the present situation, this seems be one of the major causes why State-Socialism lags behind the pace of scientific-technological developments in the »West«. But in the highly industrialized countries there is also an ingreasing debate - and ingreasing helplessness - on how to overcome threatening effects of the current path of technological evolution, which particularly affect the social welfare.
- fourthly, there surely are also harmful traditions related to State-Socialism. For instance, the almighty informal, personal interests and relations in social and economic life which obviously reduced the efficiency of the production process and had a strong influence on research and development, as well as the bureaucratic character of the whole system which acted as a strong barrier against innovation and which have left considerable remnants in today's post-state-socialist world.

Although the general situation in the countries under transformation appears to be not too bad, we must ackowledge dangerous recent developments in the sphere of science and technology. There is not only a tremendous lack of funds to maintain the achieved level of basic and production-related research but also a virtual breakdown of modern manufacturing industries in the »East«. The vanishing hopc that the loss in research capacities will soon be made up by foreign direct investments and the revival of research and development is generally not very promising – if we take a precise look at even relatively advanced countries in the »Third World«.

III

Is there no other way than to rapidly and totally submit to the predominance of Western-style science and technology and become a scientifically backward and technologically dependent group of countries?

At a first glance, nearly all results of fifty years of state-socialist scientific and technological development are inferior to the achievements of the »West«.

From this point of view, the Centraleast European countries should mobilize any effort possible to get access to the Western standards of scientific and technological knowledge. But the real problem is much more complex. Not only because of the fact that the task of combining the most sophisticated technology with the regional economic culture will remain. And the criterion of success in this field is wheather these different conditions enable at least the same levels of efficiency as those of highly industrialized Western countries. Even if the technology is available, it is by not way easy to meet the efficiency criterion and thus success is highly uncertain. The main problem is that the western path of scientific and technological evolution must nowadays be questioned in itsself. Behind the splendid success stories in the fields of production efficiency, product development and diversification - the whole colourful world of consumer goods, automatization, and globalization of Western cultural symbols - it becomes more and more clear that this kind of development is not sustainable; neither with regard to its social consequences nor concerning its ecological »side-effects«. Thus the actual challenge to transformation is not to adopt Western-designed development strategies, to add regional culture and stir but to requestion development. To overcome this challenge, the Centraleastern nations probably have some important experiences and achievements to offer. Mainly this concerns the first three points we made above. We should ingreasingly concentrate our efforts in science and technology on the solution of development problems in less advanced regions to overcome poverty as well as malnutrition and ecological devastation. In the so-called developed world as well as in the »Third World«, we must solve the unemployment problem and we must stop the waste of natural resources. This could mean reducing resource flows, to achieve more regionalization instead of accelerated globalization and this also means recombining the drive for economic efficiency with a necessary improvement in social living conditions in a sustainable manner.

However, finally we should not forget - despite of all optimism caused by the breakdown of totalitarian state-socialist regimes and the beginning of transformation to market economy and to democracy - the successful achievement of the aims of this process is by no means sure. In the »West« we also witness an ingrease in corruption and mafia structures in the economy and in the society as a whole. The Centraleastern nations with their legacies from the past are not at all immune to these dangerous tendencies.

References

Armour, L. (1995): Economics and social reality: Professor O'Neil and the problem of culture, in: International Journal of Social Economics, 22(1995)9/10/11, pp 79 - 87.

Åslund, A. (1995): How Russia Became a Market Economy, Washington D.C.

Galtung, J. (1996): On the Social Costs of Modernization. Social Disintegration, Atomie/Anomie and Social Development, in: Development and Change, 27(1996)2, pp. 379 - 413.

Köhler, G., Gore, Ch., Reich, U.-P., Ziesemer, Th. (eds.) (1996): Questioning Development. Essays on the theory, policies and practice of development interventions, Marburg.

Lapidus, G.W. (ed.) (1995): The New Russia. Troubled Transformation, Boulder.

Lavigne, M. (1995): The Economics of Transformation. From Socialist Economy to Market Economy, Houndmills.

Nederveen Pieterse, J. (1991): Dilemmas of Development Discourse: The Crisis of Developmentalism and the Comparative Method, in: Development and Change, 22(1991)1, pp. 5 - 29.

Nederveen Pieterse, J. (1994): Globalisation as Hybridization, in: International Sociology, 9(1994)2, pp. 161 - 184.

Nederveen Pieterse, J. (1995): The Cultural Turn of Development: Questions of Power, in: Development and Change, 27(1995)1, pp. 176 - 192.

North, D.C. (1990): Institutions, Institutional Change and Economic Performance, Cambridge.

Shen, R. (1994): Restructuring the Baltic Economies. Disengaging Fifty Years of Integration with the USSR, London.

POSSIBLE FORMS OF CO-OPERATION IN ENERGY POLICIES BETWEEN WESTERN AND BALTIC COUNTRIES

LUTZ MEZ
Free University of Berlin

Introduction

The energy systems in the Baltic countries were an integrated part of the Soviet centrally planned economy. After achieving political independence the enormous task of reconstruction started. In the transformation from old centralised organisations to new structures energy problems have first priority. The high energy import dependency makes the Baltic countries vulnerable, and the energy systems show severe problems, both with the supply and the demand side. These problems, due to the use of fossil and nuclear fuel in outdated equipment, must be faced now, and the environmental problems created by the energy system reinforce this. Burning fossil fuel in inefficient plants without cleaning facilities are responsible for a major part of transboundary pollutants (SO_2, NO_x, CO_2).

Baltic energy policy is at a crossing point. The energy system can continue with traditional energy planning based on large units or chose a strategy of ecological reconstruction, based on three elements:

- energy conservation;
- efficient energy supply system (e.g. combined heat and power (CHP));
- high share of local and renewable energy.

The environmental problems and the new energy policy can be solved in an integrated energy and environmental policy and by several forms of international collaboration such as joint ventures, technology support, licence transfer for advanced energy equipment and joint projects.

The Baltic Countries as Integrated Part of the Soviet Planned Economy

During the occupation of Lithuania, Latvia and Estonia, the Baltic countries were among the most industrially developed republics of the Soviet Union. The process of industrialisation during the Soviet time was setting up manufacturing industries, dependent on energy and raw materials from other Soviet republics. The plan of this economic policy was to make the Baltic countries economically dependent on the rest of the Soviet Union.

K. Prunskienė and E. Altvater (eds.), East-West Scientific Co-operation, 93–105.

Also the energy sector of the three Baltic states was reconstructed towards the necessities of the Soviet centrally planned economy.

The Soviet energy system was characterised by following institutional frame conditions:

- centralisation of the competence for legal frame setting, decision making and resource allocation in Moscow;
- hierarchical steering of the energy economy;
- symbiotic entanglement and convergence of interest of the production elite and the political-administrative elite;
- closed decision making arenas with restricted access for other interests;
- wrong mechanisms for pricing and tariffs.

This implicated for the Baltic republics, that energy sector decisions were made far away with very limited corrective possibilities of Baltic political organs and further a similar centralisation of the energy technology-structure.

This development can be demonstrated for the electricity sector. In 1961 the electricity system of the Baltic states became a part of the North-western electricity grid, a grid connection which still exists today. During siting and investment decisions the interests of the Baltic states were rather ignored, the optimisation of the electricity system was primary a case for the planning instances of the central level.

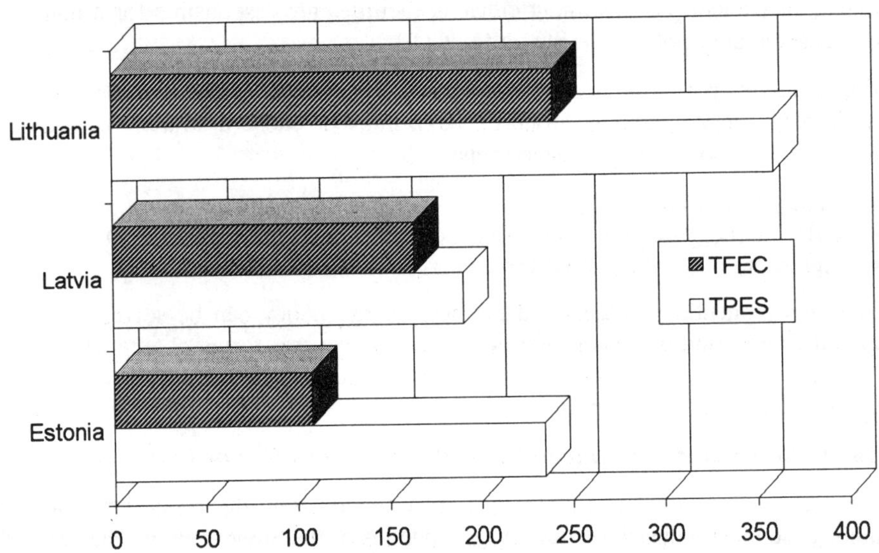

Figure 1: Primary and final energy consumption of the Baltic States 1993 in PJ

The electrical technology-structure of the Baltic States with large production units was planned for a huge region. At the same time several regional disparities for the electricity generation exists: Estonia and Lithuania are net exporters and Latvia is net importer of electricity (cf. Figure 4). A high import dependency concerning primary energy resources and a relatively high degree of energy intensity must be mentioned as further historical burdens of the since 1991 independent Baltic states (cf. Figure 1 and 2).

In the middle of the 1980s environmentalists addressing ecological problems were part of the independence movement of the Baltic countries. The environmentalists became powerful and successful. E.g. plans for the construction of a nuclear power plant (NPP) in Latvia, further hydropower stations on the Daugava river and two additional blocks of the NPP Ignalina had to be cancelled.

Political Independence and Reconstruction of the Energy Systems

The crash of the centralised Soviet planning system was in the same time the end of subordination and long distance steering of the Baltic economy. But this freedom was simultaneously a economic risk. The majority of the Baltic industry received state subsidies and had never been financially profitable.

During the first year of the restoration of independence, the productivity of state enterprises declined with more than 30% and a complete collapse of unprivatised enterprises was predicted. Further economic and social changes happened in a very limited time period, mainly three levels of changes can be distinguished:

- structural change in industry, which can be inter- and/or intra-sectoral change;
- technology change, e.g. more efficient use of resources or energy;
- behavioural change, towards sufficient people and sustainability.

A special problem was the privatisation of large state enterprises. Former state enterprises were unable to function properly because of outdated infrastructure, and to survive depended on partners in the West, willing to invest for joint ventures or to deliver know how. Co-operation with Western companies developed in several branches like machine building, textile industry and the software industry.

In the transformation from state planned economy to market economy political solutions to the bankruptcies of industrial enterprises and the liquidation of the unproductive enterprises were required. An important task for the economic policy was the co-ordination of privatisation of industry and agriculture. This included the taxation of the reprivatised enterprises. Taxing joint ventures is difficult, as the capital flow to the West should be restricted without hindering Western business to invest capital in the Baltic area.

The most important task in politics is the rebuilding of a new, loyal and uncorrupted political infrastructure. Since 1990 an own administrative structure is build up, but the Baltic states can institutionally only base on a relatively short tradition in this field during the inter World War period. The countries base on parliamentary democracy. The process of gaining political stability is dependent of historical burdens such as politicians educated in the Soviet system. And it has to deal with building up a political and administrative capacity in the Baltic states. Due to the principles of the pre Soviet time, the constitutions of the 1920s were taken as base for formulating new constitutions. For policy areas such as environmental policy no traditions were at hand, beside of environmental committees. The importance of energy questions organised the competence in this area close to the industrial and economic ministries.

An integrated environmental and energy policy administration, which would be the modern type of administration in this field, was until now not established in the Baltic states.

After five years of national sovereignty, environmental problems, the catalyst of the independence movement, are shaping the political agenda again. The conflict between energy supply and environmental protection is increasing, defined by the technological outdated power station park and the phase out of certain power stations, like the NPP Ignalina.

Environmental Problems and the Energy Systems

Since independence started, primary energy consumption declined in all Baltic states drastically. In Latvia energy consumption decreased from 1990 to 1994 with 45%, in Lithuania even with 55%, and in Estonia energy consumption in 1994 was only 57% of what it was in 1985.

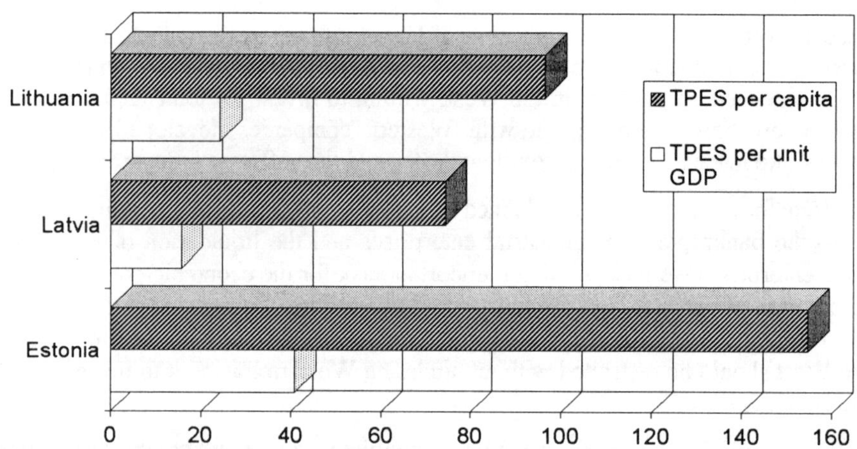

Figure 2: Specific energy consumption in the Baltic Countries in GJ

The structure of energy consumption and electricity supply differs in the Baltic states. The specific energy consumption is lowest in Latvia wit 74.4 GJ primary energy per capita. Lithuania consumes 96.5 GJ and Estonia even 154.7 GJ per capita. The same structure is valid for primary energy used per unit GDP (cf. Figure 2).

Another problem is the energy import dependency of the Baltic countries. Lithuania has almost no primary energy resources. Oil, natural gas, coal and nuclear fuel is imported. The Ignalina NPP plays a crucial role in the Lithuanian power sector. In 1995 its share of the electricity production in Lithuania was 87%. Indigenous energy resources (wood, peat, hydro and oil) represented only 7.5% in 1994. Also Latvia is a country without own resources of oil, gas and coal. The local energy resources are wood, peat and hydro. Imported energy resources constituted 78% in 1994. Estonia is importing less fossil fuels than the other Baltic stated, due to domestic oil shale production. The share of domestic fuel in total energy consumption ranges between 56% (1991) and 67% (1992). The basic producers of electricity, the big power stations in the north-east of Estonia, which are run on oil shale, are in need of immediate reconstruction, and the total efficiency of electricity production is as low as 25%. In addition shale oil is one of the most polluting energy sources. In 1994 the annual emissions per capita of SO_2 was 235 kg, giving Estonia one of the leading places in the world.

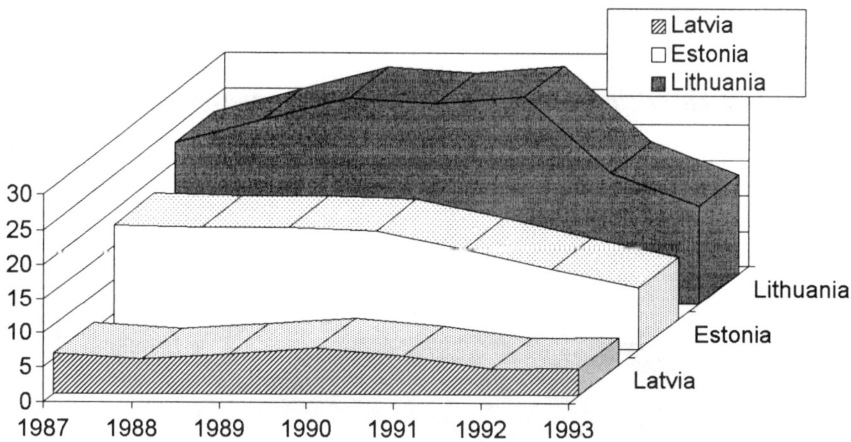

Figure 3: Electricity in the Baltic Countries 1987-1993 in TWh

Electricity production in the Baltic countries declined between 1990 and 1993 from 6.7 to 3.9 TWh in Latvia, from 29.2 to 14.1 TWh in Lithuania and from 17.2 to 9.1 TWh in Estonia (cf. Figure 3).

A similar decrease can be seen in the development of electricity consumption (cf. Figures 4 and 5). In Estonia, electricity consumption declined from 16.6 TWh (1991) to 11.4 (1993). In Lithuania and Latvia the electrical power consumption went down since 1990 from ca. 10 TWh to 7.5 respectively 6.4 TWh.

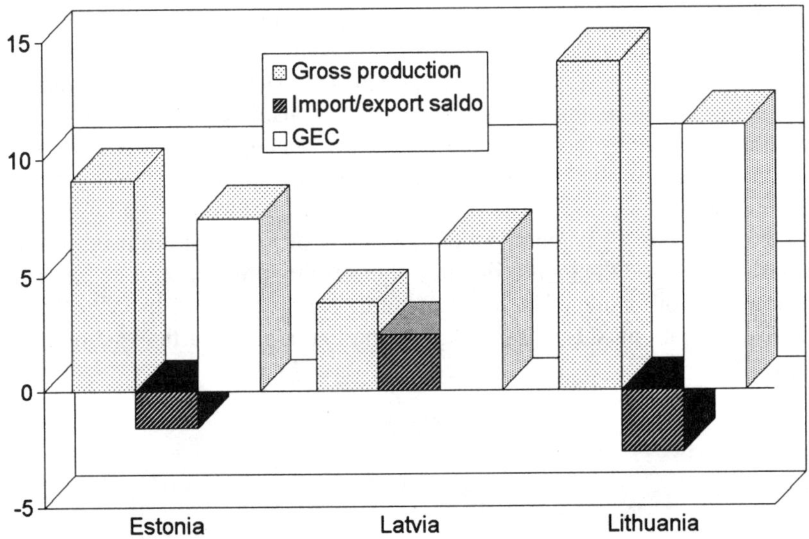

Figure 4: Gross electricity consumption in the Baltic Countries 1993 in TWh

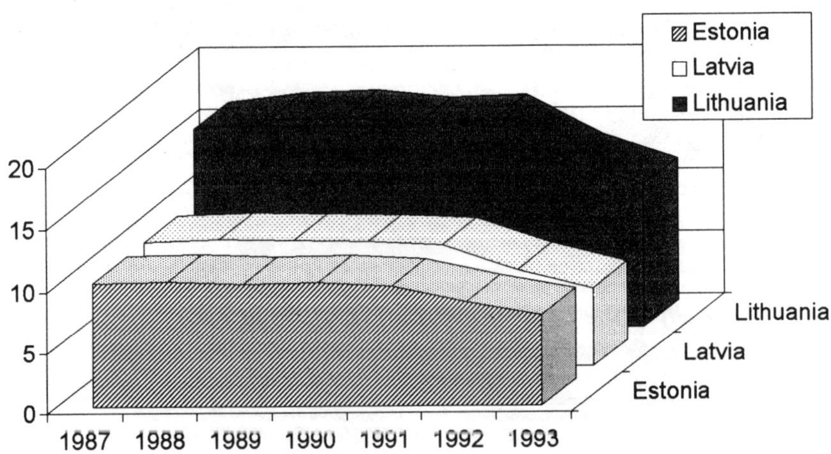

The specific electricity consumption in 1993 was relatively low in Lithuania and Latvia, while Estonia's specific electricity consumption is in the range of Western countries (cf. Figure 6)

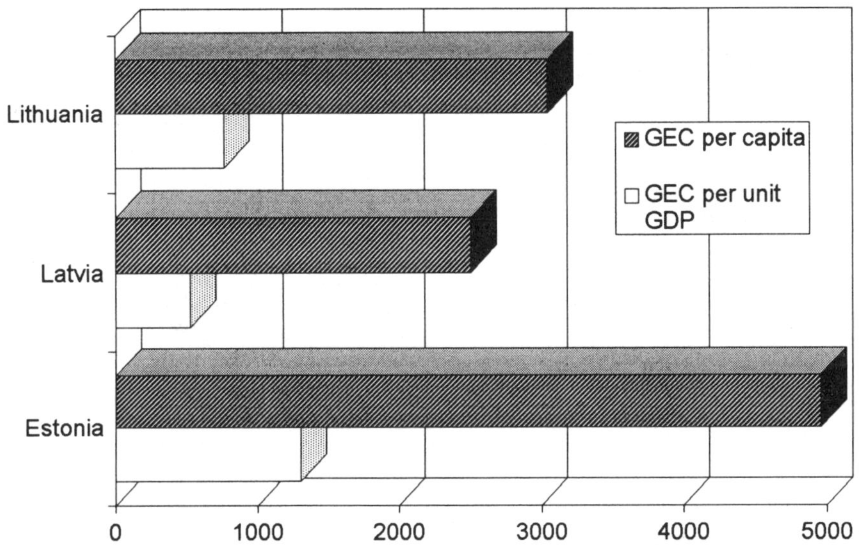

Figure 6: Specific electricity consumption in the Baltic Countries 1993 in kWh

Since the beginning of the 1990s, changes in the economic structure of the Baltic countries on the whole and in the traditional economic links and consumption of primary energy in particular had a great impact on the extent of anthropogenic pollution of the environment. Due to the decline of fuel consumption, sulphur and noxious emissions decreased significantly in the 1990s. This occurred without any special measures for environmental protection being taken. In Latvia this trend ended in 1994, when emissions increased by 19% compared to the year before. Also in the other Baltic states a trend shift appeared. The actual emission level is shown in figure 7.

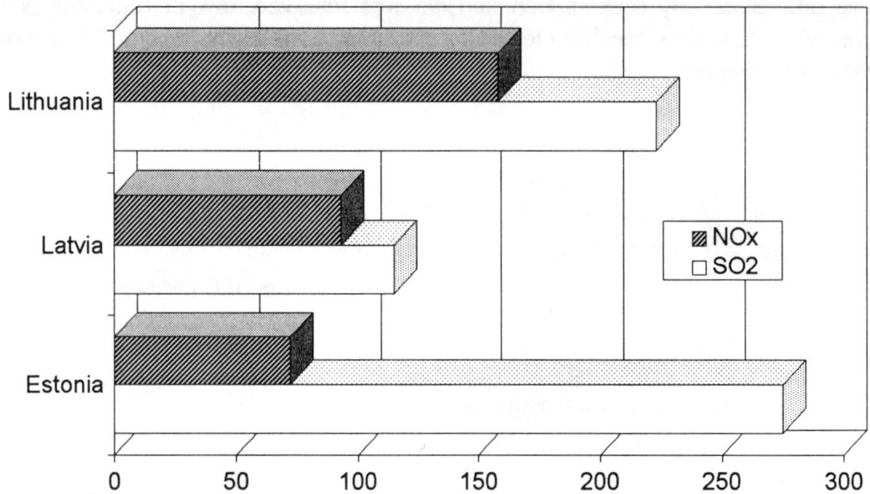

Figure 7: Emissions of SO_2 and NO_x in the Baltic Countries 1994 in 1,000

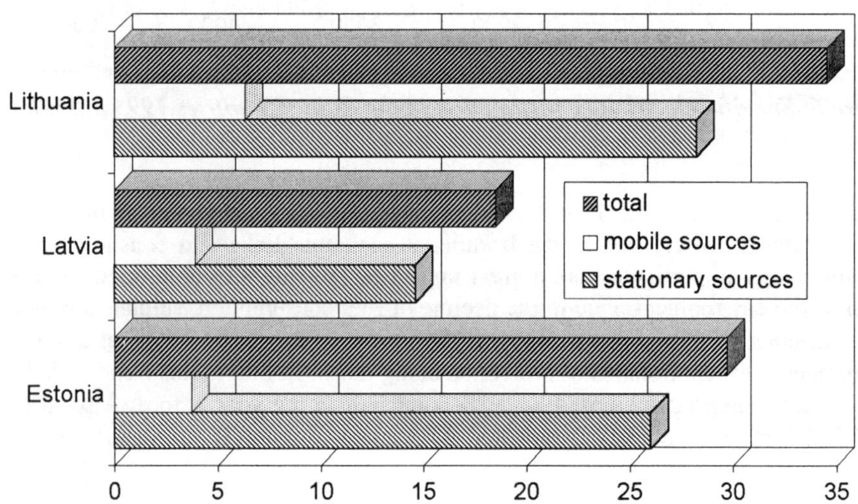

Figure 8: CO_2 emissions in the Baltic Countries 1993 in mio tonnes

Carbon dioxide emissions in the Baltic countries are in the range of 18.5 mio tonnes (Latvia) and 34.5 mio tonnes (Lithuania) and show a relatively high share of mobile sources, even if stationary sources are responsible for the major part of emissions (cf. Figure 8).

Energy Strategies

Restructuring the energy infrastructure in the Baltic countries is a decision between the traditional energy planning and an energy strategy combining future demands and ecological reconstruction. The traditional solution would be to continue in the Soviet path of economy and energy planning, with large units and paying little attention to regional and local needs. Centralised energy planning policy will continue with nuclear power, large combustion plants and imported energy as fuel. Several master plans for the Baltic countries show evidence for this energy policy option.

The inter-Baltic and regional co-operation in energy policy has to solve at least following ecological problems, exerting a pressure on the energy sector: (1) ambiant air protection, (2) resources conservation; (3) landscape preservation; (4) radiation protection and (5) climate protection.

The implementation of a demand side oriented energy policy, setting effort on the efficient use of primary and secondary energy and targeting energy use in production and private households, is asking for a fundamental shift of the energy policy. This policy has attraction from the point of cost efficiency and environmental and climate protection targets. But in all three Baltic states there exist more or less a tendency for the reproduction of traditional, supply oriented centralised big technologies. Further the option for nuclear power arrived on the political agenda of Lithuania and Estonia.

A key figure is the prognosis of the future electricity demand. In modern energy planning detailed information about the energy system is necessary. Sound decisions can only be made on the background of detailed analyses. The trend towards energy conservation increases this need. A reliable system of collecting and processing energy data must be upgraded.

The transformation of the energy resources and economy complex, the institutional change of the energy sector, the positioning of administrative and economic stake holders, national and international ones, the power structure of vertical integrated state owned energy supply monopolies and the special Baltic regulation patterns are to be seen in the light of institutional arrangements and networks of actors.

The political and economical power structure should transform into an institutional and legal frame, able to fulfil the claims for a new, demand side oriented energy policy, implementing energy and electricity savings.

There are some conflict areas of energy and environmental policy in Eastern Europe which have to be mentioned. These conflicts are connected to energy saving potentials and industrial development in general and where to allocate investment. Further the conflicts are interrelated and constitute problems as for example:

- huge energy saving potentials in industry vs. insecure future of industries;
- largest energy saving potentials vs. high investment;
- highest economic efficiency vs. small saving potentials.

Alternatives for Baltic Energy Policy - A Strategy of Ecological Reconstruction

Reflections on ecological reconstruction strategy have to consider the established priorities on the one hand and the period for their implementation on the other hand. The question of timing such measures is of vital importance.

Considering the specific restrictions identified for the Baltic countries, it is obvious that vast areas of savings potentials can only be opened up in the medium term due to their dependence on economic reforms.

On the other hand there is the necessity of implementing updates of fossil-fueled power plants in the face of local environmental pollution. Similarly, with such updates efficiency can be increased substantially. The measures for increasing the efficiency of existing power stations or for replacing old plants by highly-efficient new plants are reasonable since they will contribute to a modernisation of the existing power stations which is required anyway. Due to a lack of capital funds or technologies not available such reconstruction measures would directly compete with capital-intensive updates of nuclear power plants where the result of such modernisations are unpredictable.

In the assessment of future demand it is rather safe to assume that due to the economic restructuring, energy demand will range below the 1990 level for at least 15 years to come. Therefore plans based on a further increase of energy demand are actually no longer valid.

An important contribution to opening up intrasectoral savings potentials is identifying areas, where investments will guarantee the intended savings effects also in the medium term. A sound economy concept has to be the foundation for this. Basically "sound" energy savings can be implemented in the area of room heating.

In this connection working out energy concepts for the integrated preparation of energy management decisions is of particular importance.

First of all the existing heating stations and combined heat and power plants should be checked for possible improvements in efficiency apart from being equipped with filters. Here, however, the effects of the large savings in room heating and thus demand for long-distance energy to be expected constitute a factor of uncertainty.

If the net utilisation ratio of an existing coal power station is increased from 25 to 33%, one third more electricity can be produced with the same amount of primary energy employed.

The existing power plants will have to be equipped with filter and purification units as well. With such reconstructions a remarkable increase in efficiency of these plants can be achieved which in turn will result in a substantial reduction of CO_2-emissions.

Of course the same factor of uncertainty with respect to future demand applies to the update of power stations - once the change of economic structures shows its effects a substantial decrease in demand is to be expected.

The share of oil- and gas-fueled power stations in the Baltic states is comparatively low. Gas and steam power station built today can have an efficiency level of more than 50%. This strategy of constructing new plants would also meet the call for de-centralising energy production.

International Collaboration and Western Support

Western support programs have to consider that as a matter-of-fact capital transfer to the Baltic states cannot be refinanced.

Any transfer payments therefore have to be particularly checked for security and efficiency of the effects they are supposed to trigger.

In particular "sound" investment objects can be found if a comprehensive analysis is carried out.

Three areas of transfer are of particular importance:

(1) Technology transfer; (2) license transfer and (3) know-how and concept transfer.

Imports of technology are basically not efficient due to the "currency barrier". That is why financial transfers for acquiring technologies first of all have to concentrate on technologies with a low level of capital required. In the Baltic states energy production will only be economically reasonable if the costs of importing the required elements are in a sound relation to domestic resources. In this connection the great attractiveness of gas-fuelled combined heat and power plants can be emphasised. In the case of updates and construction of new nuclear power stations which meet Western safety requirements, this constellation is up against capital intensive systems that have to be paid in foreign currency for the most part.

Secondly transfer services have to be traded for technologies which due to the technological backlog of the Baltic states cannot be produced in the countries. In particularly this is relevant in the case of flue gas dust collectors, which, however, have to be available within a short time.

In the long term the Baltic states have to be able to rely on their domestic production of the systems required. An efficient support policy should not merely be aimed at importing e.g. energy savings lamps at economically insufficient terms but at building up an own production.

This is particularly true for the end-of-the-pipe technologies and innovative savings technologies. The foundation of Joint Ventures is not an appropriate approach if a large part of the systems produced in-country is exported or can only be bought against foreign currency. Experience in East Europe has shown that Joint Ventures are merely used for relocating production lines. Cheap technology should be available for an ecological modernisation.

A way out of this dilemma is granting East European countries licences financed by Western countries including a maximum export quota, so that a large part of the production is available for the country itself.

Co-operation with Western countries to tackle the economic and ecological problems should also include a know-how transfer with regard to behavioural changes, environmental education and so on. An important aspect of effective Western aid is political awareness in line with actual events and requirements in the Baltic states. The development of strategies which at present is merely oriented on fictitious growth rates and technological concepts is not a sound basis neither with respect to domestic nor to foreign policy. Reasonable projections of economic development as well as general economic reflections with particular importance placed on the restrictive economic situation (in particular job effect, capital built-up in-country) as well as ecological requirements should be equally considered.

Experience shows that without substantial efforts for getting across the discussions and concepts developed in West Europe, a reasonable decision making process in East Europe will hardly take place.

In order to get across these complex processes, which in addition cannot be understood by decision makers coming from East European mental and political traditions, effective instruments have to be found. By offering an energy concept agency a counter-weight can be established to the transfer of know-how mainly dominated by sales interests of Western companies.

References

Fenhann, Jörgen (ed.), Energy and Environment in Estonia, Latvia and Lithuania, Roskilde 1992

Galinis, Arvydas & Midkinis, Vaclovas, The Energy Sector in Lithuania Ten Years after the Chernobyl Disaster and its Future Development, in: Matthes & Mez (eds.) 1996, pp. 70-84

Kudrenickis, Ivars & Klavs, Gaisdis, The Latvian Power Sector Today and its Perspectives for the Future, in: Matthes & Mez (eds.) 1996, pp. 49-69

Lausmaa, Tonu, The Estonian Energy System in Transition - Problems and Solution, in: Matthes & Mez (eds.) 1996, pp. 85-97

Matthes Felix Chr., Mez, Lutz & Wanke, Andreas, Alternate Strategies for Western support of environmentally friendly energy management in Russia. Short expertise on request of Greenpeace Germany e.V., Berlin, June 1992

Matthes, Felix Chr. et al., Neue Energiepolitik für die Ukraine. Die Rolle der Energieeffizienz für eine moderne Ukraine, Kiew 14.-15. Oktober 1995

Matthes, Felix Chr. & Mez, Lutz (eds.), Electricity in Eastern Europe. 10 Years after the Chernobyl Disaster, Berlin 1996

INTERNATIONAL COOPERATION IN THE SOCIAL SCIENCES.
The Case of Western Europe and the Baltic States

ERIKA SCHWEFEL
German Social Science Infrastructure Services (GESIS)

I have come here as a representative of the German Social Science Infrastructure Services (GESIS) whose main task it is to collect and transfer information on research, literature, surveys, and data-sets from social scientists of the German speaking countries. In addition, our Berlin Branch Office is active collecting news on the social sciences in Eastern Europe and mediating contacts and information in both directions. This has helped us to gain some insight in the development of the social sciences in Eastern Europe. Until now we have concentrated on the Community of Independent States and the Visegrad countries. Therefore, from now on, it will be my job to get busy on the Baltic States and I certainly welcome this great opportunity to speak to you here and to learn from you in the course of this workshop.

My topic is international cooperation in the social sciences, especially between Western Europe and the Baltic States. My contribution is structured in three parts:

I. The importance of international cooperation in the social sciences in general and specifically in the case of Western Europe and the Baltic States.

II. The state of the art of existing cooperation in the social sciences between the Baltic States and Germany.

III. The services of GESIS in fostering East-West cooperation in the social sciences.

I. The importance of international cooperation

We are here at this conference to talk about cooperation between our countries. Our presence already expresses that we are convinced of the necessity of international cooperation in the sciences and of its helpfulness for the development of the sciences themselves but also for the development of the countries concerned. Today's societies are rapidly changing worlds in regard to knowledge, innovation, methods of conflict management, communication technology, social structure, aims and values of people. But while it is fairly easy to define what is an advance in the technical fields it is hard to do the same for the social sciences.

K. Prunskienė and E. Altvater (eds.), East-West Scientific Co-operation, 107–118.

In the social sciences, the issues are more sophisticated and often located on a more abstract level. These sciences use an analytical approach and offer new insights rather than practical solutions. What makes the social sciences so revealing is the understanding one can gain of the functioning of systems, groups, organizations, institutions, the mechanisms of power, of economic development - just to name a few issues. Since the Eastern European societies go through dramatic changes and experience, there is a growing need to understand these processes by means of social science research. This understanding is a prerequisite for developing effective practical solutions.

What can we gain from international cooperation in the social sciences? We can study topics of vital importance that cannot be treated adequately on local or national levels, they ask for international cooperation like for example environment, resources, migration. In addition:

1. Cooperation teaches humbleness, as patterns of life taken for universally valid fit only our own narrow cultural context;

2. Cooperation helps to develop a more sophisticated research by enlarging the background experience to more than one culture;

3. We learn that solutions developed and successfully implemented in one society will very likely not work in a different setting, but will need to be adjusted;

4. Cooperation in comparative research reveals problems in a new perspective and opens our view to new alternatives.

Last not least, there are profits for the individual researcher: A chance for acceptance in the specific international scientific community will lead to increasing invitations to workshops, conferences, and will give access to better publication possibilities thus maybe opening career chances.

Regarding today's situation in the Baltic States aiming at transforming their societies to integrate them to Western Europe we encounter specific problems. Do the people know what our Western societies are like? Do they understand how they function and what essential structures are their backbone? Which structures have to be built up in order to avoid chaos and allow for self-sustaining development? In addition to changing the government, permitting political parties and restructuring the economy, many social and cultural measures have to be taken. These measures can only be developed on the thorough knowledge on how the transformation goes on in the three Baltic countries, how it affects their different regions, social classes, age and gender groups. There also has to grow an awareness for the social costs and possible measures limiting these as we have learnt from our (very different) transformation experience in Germany.

International cooperation can help the Baltic social scientists to learn more about and understand the functioning of Western societies. On the other hand Western researchers interested in the Baltic States simply cannot carry out valuable and reliable research without having competent partners at the local level. Integrating these Baltic researchers in their teams and on their payrolls will also encourage Baltic researchers to stay in the world of science and not to change for better paid jobs to companies and government offices thus lessening the internal brain drain.

To understand my later arguments we have to look back briefly at the situation of the social sciences in the Baltic States during USSR-times. I will summarize for you what is reported on the situation of the social sciences before 1990[2]. Taking sociology and political science as two main branches of the social sciences we are informed that sociology flourished rather well in the 1960s in many parts of the former USSR. But it experienced difficult and narrowly censored periods in the 1970s and 1980s until Gorbachev's „glasnost" blew up the barriers to sociological research. What has been piled up from these 25 years of „limited" research is a treasure of descriptive empirical studies in fields such as youth and family research, investigations of industrial work places, time budget and leisure time studies, and elaboration on methods of empirical research. In university education, however, choosing sociology as a major subject has been possible in Estonia only since 1989. Therefore, the Baltic countries did dispose of a number of fairly well trained sociologists mainly originating from neighboring subject areas. The situation of political science has been much worse. Central research topics such as political power, functioning of political institutions, elite analysis, and interest conflicts were taboo. There existed some opinion research as top secret endeavors of communist party academies and also some very ideologised research on international relations as well as on international institutions.

The late 1980s and the early 1990s have brought about a tremendous increase of social science research in the Baltic States, especially in public opinion research with many new private institutions set up and with many activities to catch up with research as found in the respective international scientific community. However, economic problems and serious budget cuts result in new obstacles for

---

[2]Gaidys, Vladas: Social Sciences and Their Empirical Basis in Lithuania in 1984-1994, in: Heinrich Best; Ulrike Becker; Arnaud Marks (eds.): Social Sciences in Transition. Social Science Information Needs and Provision in a Changing Europe. Proceedings of a European Conference in Berlin, November 11th-13th, 1994. Informationszentrum Sozialwissenschaften, Bonn 1996 (Europe in Comparison, vol. 4), pp. 77-84.
Vetik, Raivo: Social Science Research in Estonia in the 1990s, in: Best, loc.cit., pp. 129-135.
Alisauskiene, Rasa: Lithuania, in: Hans-Dieter Klingemann, Ekkehard Mochmann, Kenneth Newton (eds.): Political Research in Eastern Europe. Proceedings of the Second Conference in Berlin, November 26th - 28th, 1993. Informationszentrum Sozialwissenschaften, Bonn/Berlin 1994, pp. 115-118.
Semanis, Einars: Latvia, in: Klingemann, loc.cit., pp. 107-114.
Vetik, Raivo: Estonia, in: Klingemann, loc.cit., pp. 71-76.

the development not only of the social sciences, but of them very decidedly. Their importance in the ongoing transformation period is being underestimated. So there is plenty of action space for international cooperation.

II. Existing cooperation between the Baltic States and Germany

Coming to our central topic of the cooperation between Western Europe and the Baltic States in the social sciences, there is a general agreement that the three Baltic States enjoy their closest international relations with the Scandinavian/Northern European countries, Germany, and the Netherlands. This seems to be natural for reasons of neighborhood and historical ties. An additional factor is the large number of Baltic exiles in all of these countries.

In the Western countries up to 1990 contemporary social science research on the Baltic States was a rather exotic endeavor. The contributions to the established journals on Eastern Europe were mainly historically oriented with some consideration of other humanities' subjects like language and literature. Looking more closely to the German language social sciences by analyzing our GESIS research project and literature databases we find very scarce incidence for a special interest in that area.

Comparing the two five-year periods 1986-1990 and 1991-1995 we find only 22 literature items and 3 research projects in the first one, but 95 and 48 in the second.

Table 1 Number of literature items (journal/reader articles, books, unpublished research reports) in the database SOLIS (Social Science Literature Information System) and number of research projects reported in the database FORIS (Social Science Research Information System)

Time period	*SOLIS documented literature items*	*FORIS reported projects*
1986-1990	22	3
1991-1995	95	48
Total	**117**	**51**

We clearly see that the interest in dealing with the Baltic States has increased impressively and this is promising - because without an interest on both sides, a fruitful cooperation is not possible. Who are the researchers that develop such an interest, where do the authors of so many articles come from? For the researchers there opened several paths. Some authors returned to their family roots, dug up language abilities and family ties, and started with field research that before had been impossible for political reasons. Others have been in the field but switch to more contemporary research topics and thus pass from history to political science. So these new chances to access information and to travel freely were realized. We are not to forget that the implosion of the communist world and the impressive struggle for independence of the Baltic peoples motivated many researchers not only in Germany to take a closer look at this region of the world. This endeavor was strongly supported and also initiated by an increasing demand from political bodies for more information about this formerly marginalised part of Europe and the change processes taking place there. These political bodies - I refer to the German experience - were not only interested in more information but also inspired by a serious desire to help the democratization process continue. This is important to our countries for our own long-term interest. It is a mixture of expecting growing markets, of having neighbors aiming at the same type of economic and social system. The German government supplied extra funds to a whole range of institutions engaged in sponsoring research in general, research cooperation, and academic exchange. Some of the more important ones are the Volkswagen Foundation, the German Academic Exchange Service, and the foundations of the political parties.

It would be great to give a complete picture of all the cooperative activities in the field of the social sciences. However, this is not possible for several good reasons: Due to our decentralized political and academic systems there is no central agency documenting such activities. People entering a foreign country on a tourist visa are not asked for the purpose of their stay. The information available has not been systematically collected and studied. It could be the object of another research project, but we were neither able to carry it out nor do we consider it necessary for an adequate discussion of the matter here.

Before I give you an account of the different types of cooperation existing, I would like to point out one very general problem that affects cooperation deeply: It is the severe language barrier. In my opinion this hardly discussed issue deserves much more attention and activities. I am convinced that the social sciences are especially depending on language because this is their only way of expression. Their study objects are largely abstract constructs, are very complex and need to be understood through verbal explanations. Many terms have been coined in specific national, regional or cultural contexts and have no meaning or a different one in other areas of the world. If this background is not known there will be many misunderstandings. This demands a thorough language instruction that includes regional studies.

Foreign languages taught in secondary schools usually cope with this problem fairly well. In the Baltic countries, however, the foreign language needed in science has changed from Russian to English. English and German (which is used much less) are third and fourth foreign languages and cannot rely on this long-term learning background. This will remain a serious problem for a decade or two unless more assistance is spent in and to the Baltic academic world to increase the English and maybe German language abilities of the staff at all levels and of the university students.

Returning to our subject of cooperation existing in the social sciences let us differentiate between distinctive types of them. I will group them according to the intensity of cooperation involved:

1. Conferences, meetings, workshops

To attend scientific gatherings is an easy way to meet colleagues, if the site is close to where you live or your expenses are being subsidized. This in mind, all kinds of meetings were encouraged and organized. Nobody has counted the number of such events, but it seems that there have been many and there still continue to be. The largest conferences seem to be the World and European Congresses of the professional associations with hundreds, even thousands of participants. Baltic participation has increased over time, counting 7 social scientists at the First European Conference for Sociology among 600 (Vienna 1992), but 35 participants in the Second European Conference for Sociology among 1000 (Budapest 1995)[3].

However, the effectiveness of getting into conversation with colleagues, of getting acquainted with formerly unknown people is generally greater at smaller workshops and meetings. The most fruitful of them - according to my experience - are those where a limited number of participants stays together in the same conference site for at least one night and take several meals together. This setting encourages informal conversations which allows for the discovery of common interests. It also helps the more timid participants and those still struggling with the foreign language to discover new starting points for joint activities that prepare a future co-operation. In addition to the already mentioned sponsoring bodies the German cultural centers in Tallinn and Vilnius, the Goethe-Institute in Riga, and many kinds of academies, adult education agencies as well as many private initiatives are active in organizing these types of meetings.[4]

2. Publication opportunities, translation or editing services for Baltic researchers, translation of important social science publications (in excerpts) into Baltic languages, exchange of publications

Looking at the authors' names of our database SOLIS entries we suspect that about 15 per cent of the 1991-1995 entries are from Baltic authors. A good number of them are articles based on conference contributions, others received publication and translation assistance as guests of German institutions.

A very useful cooperation is also the exchange of publications, although the need and interest for publications in the Baltic languages will be rather limited

[3] Newsletter „Sozialwissenschaften in Osteuropa", September 1992, p. 10. Figures were calculated from the Program and Directory of the Second European Conference for Sociology, Budapest 1995.

[4] A very comprehensive description is given in: Heinrich-Böll-Stiftung e.V., Baltischer Christlicher Studentenbund e.V.: (eds): Die Baltischen Staaten. Entwicklungskonzepte im Dialog. Handbuch der Deutsch-Baltischen Beziehungen, Köln 1995.

in Germany. The exchange will therefore have to be an almost one-way affair. With an increase of publishing activities in English this will change.

Reports from German colleagues returning from Baltic countries also state a great need for Western social science literature in general. The library programs set up in the early years after independence seem to have been only the famous drop of water on a hot stone. What is needed now are generous donations of international literature including subscriptions to the leading journals. Another serious need is noted for textbooks and other translated teaching materials. Since background knowledge of the functioning of Western political and social systems is still limited, many of the existing introductory books cannot be used, because they were written for students in the United Kingdom, the United States or Germany and take a western political socialization for granted. Therefore, a translation - if not commented - would not be useful. This demonstrates well the more complicated situation of the social sciences compared to the technical and natural sciences.

3. Exchange of personnel

There is a whole range of cooperation endeavors in this field which comprise student exchange, exchange of researchers, guest lecturers, and professors. I have neither figures on the number of people from the Baltic States who study in foreign countries nor from Germany who study in the Baltic countries. Some German doctoral students are known to spend some limited time there for empirical research. Because of the language barrier, experienced people think it is necessary to send students for a rather long period of time to enable some language integration. Short term visits are judged to be without much value except for first contacts and for visits in existing long-term cooperation programs.

The exchange of lecturers and professors takes place institutionalized and informally. Therefore there are no exact quantitative data on the amount of exchanged personnel. We can state with some certainty that the Baltic institutions are not yet flooded by foreign personnel and we know that there is more mobility from the Baltic States to the Western countries than in the other direction. This seems to be the case for various reasons that are beyond the reach of even carefully planned measures. Let me name three important ones:

1. The difference in the standard of living and of the working conditions. It is always easier for people to switch to a country where the economic situation is more comfortable.

2. Language. Students, researchers and lecturers are more likely to move to a country where they can communicate not only in very small circles. The motivation of studying a foreign language is much higher for a widely spread language than for a language spoken only in small populations, as in the case of the Baltic States.

3. Thrill for foreign experiences. The thrill for foreign experiences is still larger in the Baltic States, because of the scarce opportunities before the 1990s.

Nevertheless, also in our Western countries there is a certain number of qualified specialists who command these languages and who are motivated to move temporarily to the Baltic States. We know social scientists who are so interested in this challenging work that they are willing to learn these national languages. Those people can reach the aims of cooperation much better than those who will have to communicate in English or German - at least for the next decade. Another important argument for the teaching in the national languages is that an instructor can be more effective if he is able to get to know the audience well. From reports of German exchange professors who taught in the local languages we know that the number of students reached is tenfold. These returnees also point out to the importance of the exchange person staying in town for at least a month and better a semester or two and keeping in touch with the students. Professors who fly in once a week to give their lecture in a foreign language really leave no trace and mean a waste of money. All exchange programs profit from longer stays.[5]

The exchange of researchers in the social sciences between the Baltic States and Germany seems to be rather small so far. In a study on guest researchers from Eastern Europe in large German non-university research institutions no Baltic researchers were reported for the social sciences for the time-span 1990-1993[6]. This study found very few social scientists from Eastern Europe working in Germany temporarily, the natural and technical sciences dominated. Citing information from one of the German sponsoring bodies whose support mainly originates from federal and state budgets, 13 post-doctoral researchers in the social sciences and humanities from the Baltic States received a grant during the last years but no German chose this region in the same program.[7] Other reports

---

[5] Manfred Kerner: Politikwissenschaft in Lettland, Berlin (Freie Universität Berlin) 1995 (Materialien und Dokumente zur Friedens- und Konfliktforschung, Nr. 22).
Dirk Holtbrügge: Erfahrungen bei einer Kurzzeit-Gastdozentur an der Staatsuniversität St. Petersburg, in: Osteuropa. Zeitschrift für Gegenwartsfragen des Ostens, vol. 45, no. 11, 1995, pp. 1067ff.

[6] Hedwig Rudolph: Ex Oriente Lux? Gastwissenschaftlerinnen und Gastwissenschaftler aus Mittelosteuropa und der ehemaligen UdSSR an deutschen Forschungsinstituten, Wissenschaftszentrum Berlin für Sozialwissenschaften, Berlin 1994, discussion paper.

[7] Information from the Alexander-von-Humboldt-Stiftung, September 1996.

confirm this fact[8]. A total of about 40 Baltic researchers are estimated to stay in Germany at the time, mainly in history, language and economics.[9]

4. Research networks in international comparative research and data collection
Groups of researchers from different countries who decide on a common topic that they want to study in their respective home countries can be called research networks. Usually, there exists an agreed upon research outline and methodological approach to ensure comparable findings. Periodically, these research partners meet in real and virtual workshops and exchange their experiences. In their daily work they remain in their own country. Generally, these groups dissolve after terminating their joint publication. These networks have a number of advantages: They allow the researchers to work at home, but make them familiar with international standards. They demand only moderate language skills. Often, financial assistance can be granted, the travel costs remain rather low. The local expertise of the researchers is used, but their understanding of the problem beyond their national horizons increases.

Our database „Research in Eastern Europe" contains several such projects. In one of them Estonia is participating among 12 countries (topic: International social justice), in another Lithuania is one of four involved countries (topic: Comparative international research on external migration). The before mentioned database SOLIS lists several cooperative projects: The project with the topic "The future of industry in Central and Eastern Europe" included Lithuania among 5 other countries. A major cooperation (12 participating countries) sponsored by the German Ministry of Education, Science and Technology on „Transformation of science systems in Central and Eastern Europe" included all Baltic countries.[10]

There also exist some permanent projects like the Central and Eastern European Euro-Barometer, which was initiated in 1990 by the European Commission. It is a yearly survey that is also conducted since in the three Baltic countries. The International Social Survey Program (ISSP) has so far only included Estonia (in 1995), while the International Committee for Research into Elections and Representative Democracy (ICORE) comprises all three Baltic countries. The later ones also involve other continents and are listed here only as examples. Data files for further analysis can be supplied through GESIS.

5. Institutional cooperation, curriculum development, international ventures
The formal agreements signed by institutions of higher education are a special kind of international cooperation allowing many kinds of activities like student and

[8] EuroFaculty Newsletter, no. 6, May 1996, p. 7.

[9] An information passed by the director of the Documentation Center on Baltic Culture in Trier.

[10] Renate Mayntz, Uwe Schimank, Peter Weingart (eds.): Transformation mittel- und osteuropäischer Wissenschaftssysteme. Länderberichte, Opladen (Leske + Budrich) 1995.

faculty exchange, curriculum development, research cooperation and others. For Germany, we can clearly demonstrate that the number of such agreements (in the social sciences) with partner institutions in the Baltic States has increased from just 4 partnerships in 1987 to impressive 41 in 1996[11]. Looking at today's list of concerned academic subjects in a ranked order we notice that economics (16), education (13) and history(11) are the top three being followed by law (9). Social work (3), political science (2), sociology (1) and psychology (1) still have very little institutional attachment in the Baltic States, although as I argued before, political science and sociology could help understand the process and accompanying aspects of transformation.

In 1993, an international cooperation was started with the Euro-Faculty in Riga and its branch offices in Tartu and Vilnius. The Euro-Faculty was set up by the Baltic Council to improve higher education in economics, law and public administration in order to facilitate the aimed integration to the European Union. Another goal was to strengthen the cooperation among the Baltic universities and to coordinate the exchange of scientific personnel within the Baltic States and other countries in these academic fields. Universities from Denmark (2), Finland (2), Germany (6), Norway (2), Poland (2), and Sweden (1) organize this project. Initially, only the qualification of existing academic staff was intended. Later, the program was extended to train teaching associates and to teach students. This happened because the internal brain drain had reduced the number of young faculty members.[12]

In order to compensate for insufficient language qualifications and old-fashioned study and learning habits, English and German academic preparation courses were added to the program. The instruction of the 37 courses held in the fall of 1995 was administered in English, German, or the national languages. It is planned that in the near future the undergraduate courses can be taught by local faculty members in the national languages. With the assistance of the partner universities some junior faculty members have received additional qualification by study visits in Wesern countries. Lately, the work of the Euro-Faculty has received additional help from the Civic Education Project that is mainly financed by the

---

[11] 3 partnerships were reported from the GDR in Bärbel Last, Hans-Dieter Schaefer: Die Wissenschaftsbeziehungen der Hochschulen der ehemaligen DDR mit Osteuropa. Kompendium, Projektgruppe Hochschulforschung, Berlin 1992; 1 partnership is listed for the Federal Republic in: Westdeutsche Rektorenkonferenz: Kooperationsvereinbarungen (Partnerschaften) zwischen deutschen und ausländischen Hochschulen, Bonn 1987.
For 1996 I obtained computer printouts (dated July 1996) from the Hochschulrektorenkonferenz Bonn.

[12] Heinrich-Böll-Stiftung e.V., Baltischer Christlicher Studentenbund e.V. (eds): op. cit., pp. 91ff.

Soros Foundation[13]. So far, the number of courses offered increased every semester and so did the number of students attracted.

Euro-Faculty members report rising academic standards in economics, law and public administration at the three involved universities. To their judgment the program successfully assists the modernization and upgrading of higher education in the Baltic States.

III. The services of GESIS in fostering East-West co-operation

Because of the magnitude and diversity of activities in the social sciences, no single institution is able to store all information needed for so many academic purposes. We have set up our Branch Office in Berlin in 1992 using the language and background knowledge of colleagues from the dissolved GDR Academy of Sciences. The Berlin Branch Office serves to ease the information flow between Eastern and Western European social scientists and those bodies or individuals interested in their work. Hereby we contribute to fill the gap that the breakdown of national infrastructure services has caused in Eastern Europe. As I already pointed out in the beginning, GESIS collects and transfers information on research, literature, institutions, experts, surveys, and data-sets in the German speaking countries. These activities allow us to pass information on the social sciences in Germany to an interested audience in other countries, thus mediating mutual exchange. In Berlin we carry out a special survey on social science research projects in Eastern Europe, our third survey started a week ago. Besides, we collect various kinds of meta-information for example on Eastern European social science journals in foreign and national languages, yearbooks, handbooks, brochures, and information on research sponsoring. We search the Internet to find information supplied by scientific institutions. In addition, we have built up a net of national correspondents in various East European countries who supply us with material and report news of the national social science professional associations, past and coming conferences, and other relevant information. Let me summarize our services to Eastern and Western Europeans: We offer information consulting and retrieval services, the publication of reference books, state of the art and conference reports as well as current awareness series, and newsletters.

Our latest project involves using the Internet potential to establish a clearing house function by referencing to sources and networks on and from Eastern Europe. As a special service we supply technical support to institutions which, for technical or organizational reasons, are not yet able to access the Internet independently.

I could easily go on and inform you in detail about all our products and services. Since this would take another hour, I invite all of you who are interested

---

[13] EuroFaculty Student Bulletin, No. 4, Spring 1996, p. 1; the support consisted of 5 lecturers during the fall semester 1995 (EuroFaculty Newsletter, no.6, May 1996, p. 9).

in more details to talk to me and have a look at the sample copies and other materials that I have brought with me.

SCIENTIFIC AND TECHNICAL POTENTIAL OF THE REPUBLIC OF BELORUS AND POSSIBILITIES OF CO-OPERATION WITH THE BALTIC AND WEST EUROPEAN STATES

V. RUSAKEVICH
Deputy Prime Minister, Republic of Belorus

The republic of Belorus possesses the considerable scientific and technical potential which makes it possible to solve the main scientific and technological problems of economy and social sphere. Suffice it to say, that almost 300 organisations including 133 scientific and research, 54 design, 21 planning and research institutes, 34 higher educational institutions, 17 scientific, research and design departments at the industrial enterprises are involved in research and development.

We manage to partially finance this research from the budget. Besides, recently we actively developed works on the private basis. There are more than several hundred of them now.

More than 50,000 scientists including 1,300 doctors and more than 13,000 candidates of sciences are engaged in the field of science.

The bulk of the scientific and technical potential of the Republic of Belorus is concentrated in the productive branches of economy - industry, agriculture, transport, and communications. In these branches, 58% of all the scientists are engaged and 63% of the volume of scientific developments are carried out.

The industrial profile of the Republic is determined by mechanical engineering including automobile and agricultural machine-building, instrument-making industry, radio and microelectronics, chemical and other scientific-intensive branches. The production of mineral fertilisers and oil-chemical products has a high ratio in industry.

The enterprises which produce scientific-intensive products of military and double purpose technics such as electronic and optical instruments, means of communications and control systems, as well as transport vehicles for special use, play an important role in industry.

As the role of industry in the further development of economy will remain leading and the problems of its structural and technological restructuring are extremely important for access to external markets, we intend to strengthen the scientific and technical potential of this branch.

K. Prunskienė and E. Altvater (eds.), East-West Scientific Co-operation, 119–123.

I regards to the agro-industrial complex, scientific research is being conducted by the Academy of Agricultural Sciences, which include 15 scientific and research institutes, 8 experimental stations involving over 4,000 scientists, 19 doctors, and 663 candidates among them. All these institutions are big scientific schools in plant-growing, cattle-breeding, and soil science, which covers all the problems of agricultural production.

The scientific and technological potential in construction is accumulated in 14 organisations, which develop technologies of ceramics, thermal insulation, wall construction, covering, roofing, and binding materials and porosity fillers, and testing of construction materials.

But, frankly, in conditions of budget constraints, it is very difficult to do that. The Republic suffers great difficulties in financing fundamental science. We would like to rely on the relationship with big scientific centres in Europe in this field, more so that a number of trends in the Belorussian science is very prestigious. The developments of our scientists in laser physics, means of automatisation, design, optical electronics, chemistry, and ecology, including the measures to overcome the effects of the Chernobyl disaster.

The higher education system has a great scientific and technological potential with wide possibilities to attract talented youth to take part in research, where they can achieve flexible relationships with all the branches of the national economy and foreign partners.

Yet, we think that this potential is not used on a full-scale basis. It is likely that somewhat indifferent approaches towards the higher school of the past years still exist.

The scientific and technical activities in Belorus are based on the programs of the Cabinet of Ministers. There are 33 such programs now.

The majority of the developments are aimed at the structural adjustments in the economy, resource conservation, increasing quality and the competitiveness of the products.

Thus, under the Machine-building Program, there have been designed a number of new tractors with engines from 6 to 150 hp, and they are ready for lot production.

Under Belelectronics Program, there has been developed the element base for television technics of 5 and 6 generation, means of communications, consumer radio-systems, elements for microelectronics frequency appliances and optical and electronic technics including elements for satellite and cable television, as well as for complicated technological equipment. Only Minsk research and production corporation “Integral” produces more than 200 types of new products, 90% of which are exported to various countries including the USA.

The protection of scientific and technical potential made it possible to avoid the destructive processes in economy under the severe economic crisis during the past few years.

This year, we managed to stabilise the economic situation in the Republic and to achieve, though not very high, the growth of the GDP. The export increased by 21%.

Next Year, we anticipate the growth of the GDP by not less than 5%. We clearly understand that it will be difficult to move forward using the existing technological base. Therefore, we pay special attention to the problems of improvement of investment policy and modernising the scientific and technical potential in Belorus.

According to our estimates, in order to modernise production capacities, it is necessary to invest in the economy not less than 50 billion USD in the next ten years. At present, investments from all sources amount to 1.2 - 1.5 billion USD, which is far from sufficient. The burden of obsolete technologies, low competitiveness of a number of products makes it difficult to integrate the Republic into the world economy.

The radical restructuring oriented with the use of intellectual resources and development of intensive technological production is the way out from the current situation.

The adoption of the national program to attract investment was one of the actions to solve these problems and to maximise internal mobilisation and external resources.

Certainly under the economic crisis connected with the restructuring of the economy and inflationary processes, we feel the deficit of investments. Suffice it to say, that comparable to the level of 1991, the centralised investments decreased by 5. The average annual foreign investment of roughly 500 million USD is insufficient, in comparison to the market needs.

Direct foreign investments in Belorus are made mainly through joint-ventures. It seems that their creation is very promising, together with the credits. Up to now, there are 2,000 joint-ventures and over 800 foreign joint-ventures: 59% in trade, 33% in industry. The total output is as of yet, insufficient: approximately 2% of the total industrial output of the Republic. We think that one of the most important priorities of the international co-operation is the expansion of the all-round trade, scientific and technical co-operation with the Baltic and West European States.

It is stipulated, not only by the geographical proximity, but also by the old partnerships.

I am pleased to note, that the trade among our states is increasing. So, the trade with Lithuania in 1995, in comparison to 1994, grew by 5 times, and amounted to 231 million USD, with Latvia - 3.3 times, with Estonia - 2.4 times.

Now, more than 70 joint ventures and 42 investments with Lithuanian investment are registered in the Republic of Belorus, and at the same time, there are 175 joint-ventures and Belorussian owned enterprises in the Republic of Lithuania. The total stock capital of the enterprises from both sides amounts to 4 million USD.

We have 54 joint ventures with the stock capital of 23 million USD with Latvia, and 1 million USD with Estonia.

The scope of activities of the joint-ventures is rather broad, from rendering services to producing industrial and technical products. For example, the joint-venture "Bellitelectro" produces electrical equipment and consumer goods, and is engaged in construction. The joint-venture "Transbalt" renders transport services, "LoSVer" is specialised in consulting, management, and production of consumer goods.

The creation of an adequate contractual and legal base will contribute to the further development of our co-operation. The significant steps in this direction have already been made. At the end of 1994, the Republic of Belorus and the Republic of Lithuania concluded the Agreement on trade and economic co-operation, stipulating the creation of the Belorussian-Lithuanian Commission. In our opinion, the work of this Commission is very fruitful.

I would like to stress, that we have established very good business relations with Lithuania at the highest of government levels. Our co-operation with the current Prime Minister of Lithuania is very fruitful.

We believe, that at present, we have almost co-ordinated our positions in regards to the Agreement with Lithuania on promotion and protection of investments, and we are ready to co-ordinate the final draft of such an agreement in the very near future.

The joint-stock company Litagro opened the permanent fair of the Belorussian agricultural machinery and equipment, this spring in Kaunas. It also organised the workshop on these problems. Both sides expressed great interest to develop the all-round scientific and technical co-operation, and they managed to overcome the deadlock.

In 1994, we also signed the Agreement on the principles of trade and economic co-operation with Latvia, and at present, we are practically ready to sign an agreement on promotion and protection of investments. We are finishing the preparational work to sign similar agreements with Estonia.

Unfortunately, it seems as though the possibilities of our countries are not being fully exploited. There are no essentially long-term projects.

I would like to focus on our co-operation with Germany. First of all, we have established the necessary legal base. We signed a number of agreements between the Republic of Belorus and the Federal Republic of Germany on the development of the large-scale co-operation in economy, industry, science and technology. Last year, the trade between our countries increased by 53%, in comparison to 1994. Export increased 81%.

Germany takes first place among the Western States as a trade partner of Belorus. We have created 290 joint-ventures (Belorussian-German), with the stock capital of 230 million USD. Among them, is "Opitek", which is specialised in the production of computer technology, materials for magnetic disks, and other technological equipment.

The volume of export credits from Germany is 520 million DM, or almost one fourth of all credits to the Republic.

The technological equipment for chemical and oil-chemical, regional, light industries, for agricultural product processing, medical equipment, and other goods are delivered and used in Belorus. More than 40 important projects have been financed.

The governments of the Republic of Belorus and FRG founded the Belorussian-German Council of Economic Co-operation, to co-ordinate and develop economic ties. The main directions of the Council working groups are as follows:

- agricultural and processing of agricultural products (credit lines "Hermes" with the total amount of 115.5 million DM, to buy technological equipment and spare parts for the agro-industrial complex of Belorus);
- chemical and pharmaceutical industry, production of equipment for the chemical complex (credit lines "Hermes" with the total amount of 132.5 million DM, to buy equipment for production of polyethylene gas pipes, reconstruction of the Mozyr refinery);
- machine building and automobile industry (the joint production of axles and suspensions for saddle trailers at "Belauto MAZ", together with "Otto Sauer Achsenfabrik", the production of universal milling machine tools at Gomel machine-tool plant together with Hekkert", the production of window casements at Baranovichi automatic lines plant together with "Roto Frank");
- electronics, instrument-making industry, conversion (Belorussian Scientific and Industrial Association sent the projects for lot production of municipal electric transport, production of lead-acid batteries, digital automatic telephone stations for 100,000 numbers, to the German side);
- construction, production of construction materials (projects for maintenance and modernisation of the housing stock, private housing construction, transformation of big state construction enterprises into private construction firms);
- banking (a number of workshops, with more than 70 participants from 46 banks and ministries);
- insurance (the German side provided 3.5 million DM to create an insurance training centre).

In 1996, the framework agreement on a long-term co-operation was signed with the joint-stock company Siemens.

Certainly, it is difficult to cover, in such a brief presentation, a diversity of forms of co-operation with the Baltic and west European states. But, we believe that it is necessary to act more decisively. This will inevitably enrich us, will create possibilities for young market-oriented states to acquire the necessary experience and, in the long run, find a decent place in our common European house.

UTILIZATION OF THE LATEST TECHNOLOGIES FOR WORKING OUT METHODS OF INFORMATION WARFARE AND ITS POSSIBLE CONSEQUENCES

V. D. NOZDRATCHOV

Today one can watch a process of information technologies development which takes place all over the world. Thus, these technologies and information become one of the strategic resources. Due to this tendency there appear new forms of struggle typical of the international level as well. Though this field lacks fixed definitions and terms many experts call a complex of methods and means used in this struggle "information warfare". The notion "information weapons" usually embraces methods and means which allow (a) to impair or put out of order telecommunication and computer systems, blocks ensuring "intellectual abilities" of arms, etc., (b) to steal, distort or wipe out information, (c) to cease or to limit an access to it, (d) to use it for manipulating behaviour and a state of separate individuals or large groups of people.

The peculiarities of information warfare (IW) are as follows:

- it can be carried and is carried on according to the principle "everybody against everybody" with temporary unification of some groups, countries with quite certain purposes (this item concerns mainly peace-time);
- IW can be carried and is carried on constantly (the only thing which changes with a passage of time is its methods and forms);
- IW can be carried on by separate groups of individuals with great financial possibilities (using proper strategy and tactics they can bring even big countries to their knees or manipulate them);
- a good possibility to conduct secret operations;
- a rather long latent period which passes before revealing the results of application of IW means;
- an extremely high subversive potential of specialised groups (it can be compared with use of portable nuclear charges and in some cases with an effect caused by means of mass destruction);
- super-short periods of operations (fractions of seconds) which exclude timely detection and organisation of counteraction;
- a possibility of counteraction is as a rule connected only with forestalling measures;
- if IW is lost by an enemy then it is lost for good and for ever.

K. Prunskienė and E. Altvater (eds.), East-West Scientific Co-operation, 125–128.

There are some other peculiarities which will not be discussed in the present report.

Methods and means of IW can be used with the following purposes:

- impairment of systems controlling intellectual weapon, other combat means, troops as well as systems of strategic operation and support of decision-making;
- disorganisation of activity of banks and administrative structures of a state, region or settlement;
- impairment of technologic cycles leading to technogenic accidents or catastrophes;
- formation of certain opinions, behavioural stereotypes in large groups of people;
- achieving collective altered states of consciousness (ASC) which help to manipulate behaviour and a psycho-physiological state of people.

At present the USA carries out much work aimed at elaboration of methods, means and strategy of IW which can become an element of a new doctrine known as "Force Projection". This task is entrusted to a special unit of the Ministry of Defence (the Defence Information Systems Agency - DISA)). At the same time there appear special services in the Armed Forces (the Navy Information Warfare Activity or NIWA in the Navy and the AF Information Warfare Centre or AF IWC in the Air Force).

The scale of this work can be estimated on the basis of the following fact: not less than 1,000,000,000 US dollars was spent on programmes in 1996; a volume of financing will reach 2,000,000,000 US dollars in the nearest future.

It appears that a decision on organisation of the National Information Infrastructure and Global Information Structure based on Internet was made after the Joint Security Commission had raised a question about an imperfect and ineffective defence of the USA against means of IW. Thus, it is quite evident that this country makes every effort to protect its information space and to carry on global IW with every state linked to Internet.

The operation called "A Storm In a Desert" was an expressive illustration of methods and means of IW. Evidently, its aim was to convince US legislators of the necessity to spend considerable sums of tax-payers' money on research in the field of IW. It is confirmed by the choice of the enemy on which the demonstration version was tested and which was incapable of resisting the US military might irrespective of a rather primitive set of methods and means used mainly for producing an impression on the Americans whose educational level is not very high and who are easily affected by mass media. Analysing reactions of the American society one can easily understand the attempts of the USA to instil the American culture, way of thinking and dependence on mass media in Europe where the cultural level is higher both from qualitative and quantitative points of view. It will allow to carry on effective operations in the CIS, countries of Eastern Europe and allied states. However, large-scale work in the field of IW stimulates development of both methods of counteraction and means of attack. Japan is a bright example of it. This country is going to create a national information infrastructure of a higher technological level in comparison with that of the USA by 2015-2020. One of its basic principles is maximum protection of communication channels. This problem can be solved by application of fibre optic lines reaching primary links of communication system hierarchy.

Thus, the leading countries of the world do their best to master new technologies and physical principles for creation of communication systems and information processing which will be protected from external effects.

What effects are supposed to be fought with as far as IW is concerned?

The most important fields whose quick development is paid great attention to are:

- super-broad band electrodynamics where high-capacity generators imitating electromagnetic radiation of a nuclear explosion are actively replaced by portable generators and devices which generate pulsed electromagnetic fields after an explosion; they cause directed and all-round damage of radioelectronic equipment on limited territories;
- super-fast-acting special processors which allow to decode signals controlling an operation with the purpose of providing systemic jamming;
- mastering new fields of mathematics as an algorithmic base for creating viruses of new generations and working out methods which make it possible to "break into" information systems of an enemy.

The work at principles, methods and means of IW for manipulating behaviour of a man and his state is connected with applying the most sophisticated complex of investigations. Understanding of mechanisms responsible for formation of altered states of consciousness is a key to this field. Absence of a well-thought-out conception results in an empirical approach to the problem; every researcher thinks that he and only he is on the right track. A wide spectrum of methods and means was and is used for affecting a person (pharmacology, weak physical fields, hypnosis, various types of suggestion, subliminal stimuli, etc.). Most of them are only a base for working out quick and delayed methods of information manipulation. A counteraction arouses great interest for transition from an electron carrier to a photon one. It will demand a considerable increase of an energetic potential of means which cause an effect.

As for electromagnetic signals one can observe active transition to higher frequencies, noise-like signals and development of super-fast-acting elements for protection of receiving aerials. There is a growing number of investigations devoted to use of diamond and diamond-like backing for heat-dissipation, nitrogen-free diamonds for production of semiconductor devices of a new generation. Modern achievements in the field of high-temperature superconductivity serve a ground for development of receiving protective devices. One more sphere of researchers' interest is duplicating systems based on stations of meteor communication.

The activity in the above-mentioned fields is accompanied by a gradual increase of the density of transmitted, processed and circulating information which approaches the limits of operational possibilities of new systems and makes them vulnerable to some physical phenomena which need further studying. This process can be rather long and paradoxical as it may seem it can lead to negative consequences in those countries where the discussed methods and means are worked out and introduced. Utilisation of means of communication and detection based on noise-like signals with a super-broad range, transition to a millimetre wave band will cause a considerable growth of a number of patients with functional disorders of the CNS and especially of the vegetative nervous system as well as with skin diseases, different types of allergy and as a result immuno-deficit.

This problem can be subjected to further discussion but it is necessary to pay attention to a character of conflicts taking place in the world. Global wars appear to become a thing of the past as they are too destructive and unprofitable even for a winner. They are replaced by local conflicts and various kinds of guerrilla wars. Their destructive potential is not very high. However, a guerrilla war with use of high technology methods will cause the greatest damage to those countries where a structure of communication and automation of processes is developed to its maximum level. In this connection it would be sensible to think of joint work directed at protection of information spaces and investigation of possible threatening factors. This work can be carried out by all developed countries of the Eastern and Western Hemispheres. It will make these states trust each other, reduce non-productive expenses and increase the level of safety as far as high technology terrorism is concerned.

POSSIBLE TRENDS OF TERRORISM DEVELOPMENT CAUSED BY SPREAD OF NON-FATAL WEAPONS AND TECHNOLOGIES OF DOUBLE APPLICATION: PROBLEMS OF JOINT CONTROL

V. D. NOZDRATCHOV

Modern terrorism acquires new features and is characterised by gradual improvement of its technical support with a tendency of transition from individual acts to technological and economic sabotage. This tendency is conditioned by closer interaction f ideological terrorists and organised criminal groups which are striving for an increase of their technical potential rather than fire-arms.

Breakdown of the USSR and abolition of the Warsaw Pact became an extremely favourable condition which had a great effect on the development of the above-mentioned tendency. Highly qualified specialists of secret services and branch research establishments who possess skills of operation work, an experience of manufacturing special products and tactics of their application were thrown into the street.

A considerable part of them became embittered and got out of control. The majority of these specialists was immediately engaged by various commercial organisations which are fully dependent on criminal clans owing to peculiarities of privatisation and property redistribution.

Ethnic and confessional conflicts aggravated this situation as ideas of separatism and nationalism are often used by criminal clans for masking their real purposes (avoidance of being brought to justice, securing transit and its control by a clan, etc.). These conflicts are a good "nutrient medium" for working out a platform and ideology of terrorism. They abet formation of groups of fanatics within the limits of a religious or nationalistic doctrine. Thus, the circle is closed. The result is fusion of the experience and knowledge of the former specialists of defence enterprises and secret services, capitals of big criminal clans, ideas and aspirations of fanatics. Internationalisation of criminal groups widens access to modern technologies and special means. An unstable situation typical of a vast geopolitical space makes the whole states having their own interests in post-Soviet and post-socialist countries take part in the game. As a rule it happens with the help of secret services which are prone to apply modern achievements of science and technique for receiving a maximum effect with minimum efforts.

Unfortunately, big cities become main targets of terrorists more and more often as it is much easier to carry out all preparatory work there and to organise retreat after an operation. Complex municipal services as they are often bear

K. Prunskienė and E. Altvater (eds.), East-West Scientific Co-operation, 129–132.

within themselves some elements of danger (gas pipe-lines, electric supply networks, chlorine depositories, etc.). The majority of big cities appeared due to industry development and as a rule they have several enterprises characterised by an increased risk on their territory. There are experimental nuclear reactors in many big cities of the world; atomic electric power stations, ballistic rocket silos, etc. are located in close vicinity to some of them.

Thus, post-Soviet states have become a rather dangerous supplier of qualified personnel, products and technologies providing both their own and international organised criminal groups and terrorist organisations with them.

Today such states as Iraq, Iran, Libya demonstrate a great interest in specialists and are in search of them. It concerns not only professionals in the field of nuclear weapons. Some international clans of organised crime try to do it through dummy companies. This process is quite reasonable. For example, if such two products as a system of protection from high precision weapons and war-heads to rockets with a device generating pulsed electromagnetic fields after an explosion were supplied to Iraq it could result in great losses of the USA both in manpower and materiel (including aircraft carriers) and lead to unpredictable development of the conflict. It is possible to produce these articles (for example, in Russia) at several hundreds of enterprises in the presence of 2-3 specialists. The same can be done in Iraq in case of supplying small amounts of quite harmless radioelectronic components.

Unfortunately, collapse of the system of State and military secret protection in the post-Soviet countries had a great negative effect. It not only increased the flow of secret technical information transmitted to the West and being important for leading companies of developed countries but also aroused an intense interest of criminal clans in achievements in the field of secret research.

Articles published in some popular editions and devoted to devices generating pulsed electromagnetic fields after an explosion, use of pyrotechnic lasers in the process of creation of non-fatal weapons immediately attracted attention of criminal groups to specialists in this field. The monograph written by several authors and entitled “A White Book of Russian Secret Services” gave much additional information about means of super-broad band pulsed electrodynamics and tactics of their use.

The facts mentioned above allow to come to the conclusion that it is sensible to establish a moratorium on spreading information on some fields of research and specific products which present a serious danger when used by terrorists through mass media and international computer networks. Such an approach proves its worth if to take into account possible consequences of usage of technologies and means under discussion by terrorist organisations. These consequences are incomparable with the results of application of devices described, for example, in “A Terrorist’s Cookbook”.

Adoption of such a moratorium will be complicated by absence of due control on the part of the authorities of the post-Soviet countries. Moreover, such new states as the Ukraine and Kazakhstan are ready to sell anything in spite of declaring their adherence to the idea of supervision. As for Russia the situation is more or less under control. Though the majority of specialists lead a miserable existence they want to hear nothing of contacts with the West considering that it is to blame for the USSR breakdown. It should be emphasised that many specialists are ready to hand over their developments just to extremist organisations and think that it will be their revenge on western countries.

These attitudes are formed against the background of certain difficulties in exercising control over arms development in the light of existing legislation to say nothing of technologies and products of double use. For example, spread of green lasers is under control in Russia (according to the generally accepted opinion use of such powerful lasers makes it possible to blind manpower). Nevertheless, they are sold within the country without any restrictions and there are a lot of uncontrolled channels of their transfer abroad. At the same time hardly anyone ponders over the fact that the same blinding effect can be obtained with a wavelength range which is not optimum from the point of view of energy which causes damage. It can be observed in other ranges due to applying certain parameters of radiation; at the same time amounts of spent energy are considerably smaller (it particularly concerns a power source). This makes such systems even more dangerous due to a higher risk of their secret use and absence of any control over spreading elements necessary for work in these ranges.

Tactics chosen by the Chechen opposition is one of the formidable trends of terrorism development. The experience of the Chechen conflict has shown that tactics of simultaneous multiple attacks , a seizure of several objects in big cities present a serious problem for antiterrorist units. If products based on double technologies, their elements or means from the arsenal of non-fatal weapons are used by terrorists in such conditions all efforts aimed at suppression of acts of terrorism can come to nothing. This character of terrorist activity conditions the necessity to resort both to elements of military operations and elements of police special operations. The police is not ready for the first type of operations and army units for the second one. Use of generators of super-short pulses, laser blinding systems, weak electromagnetic fields with a specially formed acoustic background, aerosols with super-tiny particles of some substances, etc. will lead to such a phenomenon as foci of uncontrolled panic among the population. Thus, there appears an extremely powerful hindering factor causing a failure of all types of communication, impairment of a psycho-physiological state of personnel of special units and other effects which will result in mass casualties and great economic losses.

Unfortunately, the activity of criminal groups in this area is much higher than that of the bodies responsible for preventing their access to the above-mentioned technologies and means as well as for working out methods and ways of counteraction.

Registered cases of use of soporific gases in trains with the purpose of robbery and hypnotic methods for recruiting women to brothels, experiments on a combined effect of drugs, weak electromagnetic fields and rhythmic opto-acoustic stimuli carried out by private groups of "researchers" are very unfavourable signs.

It means that all these methods can be used by terrorists perhaps tomorrow. They harbour one more danger: secret development of a situation. Thus, an experienced observer including specialists of antiterrorist bodies can fail to notice characteristic signs even of an active phase of an operation and loses an opportunity at least to raise the alarm. Moreover, these specialists can be involved in a planned act of terrorism when methods aimed at achieving collective altered states of consciousness are used.

The information presented above proves that it is expedient to exercise many-sided control over production of systems based on technologies of double application. There is good reason for organising joint production. In order to observe parity it should be done in the third countries. We have already mentioned several times that Lithuania is an ideal place for producing such systems. Its advantages are as follows:

- a high level of industrial efficiency close to the western one;
- a considerable number of specialists from the former military-industrial complex of the USSR who participated in development of elements for systems and technologies under discussion;
- absence of a language barrier with eastern countries and its gradual disappearance in relation to western states;
- presence of enterprises which can easily ensure mounting and operational development of products at the first stage of co-operation;
- an advantageous geographical position which makes it possible to minimise transport expenses;
- other factors.

Realisation of such a project is useful as it will help to eliminate problems which appear as a result of NATO expansion to the East.

Creation of new working places in Lithuania will stabilise the economic situation and strengthen its international status.

SCIENCE AND TECHNOLOGICAL POLICY IN THE UKRAINE: EUROPEAN AND REGIONAL ASPECTS

NYKOLA BALTAZHI
Embassy Counsellor of the Ukraine to the Republic of Lithuania

The leading feature of the research and technological situation in many newly independent states that chose the way of radical changes in political and economic structures and gradual insertion into the European community of countries and people, Ukraine among them, is persistent deterioration of scientific and technological potential. Ever reducing number of researchers and critical decline in the welfare of the scientific personnel rapidly antiquate both moral and physical material and technical basis of our science. Entire research groups, schools and institutes that were maintained and preserved over decades die out. Research productivity sharply dropped. The state cast into deep economic and financial crisis does not reciprocate the impulses originating in society for the needs of scientific and technological renovation.

Abbreviating the amount of complete research is a logical extension of the existing destructive tendencies in the areas of science and technology. In 1996 there were 179 thousand people employed in the research institutions which is two times less than in 1989. On the other hand, by the estimations carried out by the Scientific and Technological Potential Research Centre and also by the past observations of the F. Dobron National Academy of Sciences of the Ukraine the amount of people actually engaged in research activities does not normally exceed 10 percent of those employed. I do not wish to argue the opinion of many Ukrainian and foreign analysts that the above tendency is only a reflection of a natural process relinquishing from scientific and technological areas an unnecessary ballast yet it is important to mark here that the tendency if pursued will inevitably lead to ruin of all tolerable standards (i.e. violation of the so called critical mass of scientists). This, in our opinion, will put at stake the scientific and technological sovereignty of our country.

Therefore, it is not unexpected that scientologists and technological politologists share the opinion of the country's potential inability to meet economic needs in the transitional period. A striking example of the above statement is Chernobyl, whose technical needs are being supplied up to 0.9 percent. In fact, these are only the expenses bound to retain the existing level for the maintenance of the present situation without further improvement. At the same time, it is widely known that in competitive market economy new product generation processes use from 3 to 10 percent of research data. This does not concern defence area where the above

K. Prunskienė and E. Altvater (eds.), East-West Scientific Co-operation, 133–139.

number in the former Soviet Union was 33 percent and it is being stable kept at about the same level in USA.

Sharp drop in research and education prestige as well as decreased remuneration for intellectual labour caused massive withdrawal of scientists and engineers from the scientific and technological area, on the one hand, and sharply diminished influx of the young researchers, on the other. Another major reason that started erosion processes in the scientific potential is unbounded "brain drain". According to the existing statistics, in the period from 1988-1992, 219,500 people left the Ukraine. On the random count basis, specialists with university diplomas previously employed in the scientific and technological sphere make 30 percent of the above number. 1 percent of these are doctors and candidates of the sciences. In pursuit of their research interests 600 specialists from the National Academy of Sciences of the Ukraine left the country in 1995. The highest preference for the research immigration falls on Israel (75,3 percent), USA (14,5 percent) and Germany (4,2 percent).

More disturbing is the emigration of entire research groups from the unique research institutions from the Ukraine and other donor countries. This, at best, causes freezing of certain research areas. Growth of a mature well functioning research institute takes minimum ten years. It takes about 25-30 years for creation a research school. Thus the present "brain drain" and research crisis will become noticeable in full measure already in the 21st century.

Five reform years showed that expectations for quick and automatic adjustment to market structures and philanthropic attitudes had no basis. Consequences of this have been witnessed in the developmental processes of other countries exposed to market economy for decades and yet never entering the community of economic super powers. Only the countries that took meticulously pre-calculated and well focused technical and research policy steps in combination with the established market structures reached noticeable economic results. In this connection Israel's economic achievements come to mind most readily. Not long ago it counted as a developing country while now it figures among top 10 world's best developed industrial states. Israel occupies the leading position in the world by the number of scientific publications per capita as well as the number of patented research results per researcher. The university graduates percentage among working population exceeds 25 percent while in USA this number is 16 percent. This country manifests the highest rate of development among all OCEP countries (6-8 percent per year). This situation has been made possible due to the influx of intellectual force from the former Soviet Union.

The process of reforming Ukrainian science from "peripheral" as it previously was in the Soviet Union, i.e. one fifth of all research potential, into a sovereign scientific system as well as maintenance of the sought sovereignty influenced on the background of new international scientific and technological contacts is adequate. Reduction of all these contacts was the first material expression of the quandary.

In the existing situation Ukraine as a large European state with numerous historical and intellectual traditions and schools stepped along the path of gradual integration into European structures and institutions. European orientation was decided by our country including its regional policy as well. The above orientation found its expression in the participation in numerous research organisations financed by international community. This gives Ukrainian science the opportunity to maintain stable status in the areas of fundamental research as well as to secure working places for the specialists of highest qualification and, what is of paramount importance, through co-operation with foreign companies to penetrate world market with world class scientific and technological output.

Speaking about Western partners their interest primarily lies in the Ukrainian priority research areas where we have amassed huge experience and developed well acclaimed schools of research (material sciences, space and aero-technology, welding equipment, shipbuilding, newest biochemical technologies, modern informatics etc.). Western support is also directed towards preventing undesired transfer of defence technologies into the possession of antagonistic regimes.

The latter motive played the decisive role in creating the Ukrainian Science and Technology Centre in October 1993 in Kiev. The Centre became intergovernmental organisation financed by USA, Canada, Sweden and Ukraine. The main objective of the Centre is preventing military technologies and expertise dissemination by supporting Ukrainian researchers and channelling their work along peaceful needs, i.e. maintaining research focused on conversion. During its second meeting USTC in May 1996 approved of 37 new research projects from the proposed 63; which makes 60 percent and by international standards 10 to 15 percent of approvals suffice for positive evaluation of a centre activity - it's a great result. The Institute of Nuclear Research has received the best estimation for its project.

Japan, other countries and European Union declared their intention to join the Centre offering significant financial support. The sum that went to financing all the approved projects reached 4.4 million US dollars. This means that the USTC was able to give research grants to approximately one thousand qualified researchers and engineers.

Another fruitful example of co-operation with the Western partners is founding of "Paton-Prite and Whitney" research centre on the basis of E.O.Patton's Electric Welding Institute. A Ukrainian partner in this venture is the American Concern "United Technology," dealing in aerospace technology. Recent years saw a sharp rise in space programmes costs and even USA restrains from starting them on their own means and resources. Therefore USA, , together with the Ukrainian Paton Institute's technologies and expertise, world leaders in the given area will be able to develop quick advance in the area of cosmic research. Ukraine has benefited from this by obtaining the possibility to maintain its powerful scientific and technological potential in space industry and space programme market. Further impetus to Ukraine's interests in participating in as many similar project as possible was given in April 1996, when through the assistance of the National Space Committee and the

National Academy of Sciences Space Research Institute were founded. By expert estimation, the activity of this Centre during the first 10 years will bring 7 billion US dollar's worth output.

Analysing multilateral co-operation in science and technology with European partners, it is important to mark the Ukraine's active participation in different associations, initiatives and European Union programmes for post-communist countries. One could mention here INTAS, TASIS, COPERNICUS, PECO and others. The programmes help to finance priority research areas and keep up the level of technological potential in general. Thus, after INTAS was announced in operation, from the advanced 500 research projects 30 percent came from Ukraine and about 75 percent of these grants fell to National Academy of Sciences of the Ukraine projects. Financial support that went to the Ukrainian joint projects exceeded the total sum of 13 million ECU.

It is of crucial importance for Ukraine to become a member of NATO programmes together with other EC countries. These programmes provide the basis for projects in the sphere of disarmament technologies, high technologies, human resource administration and environmental protection. The programme works through the following mechanisms in the areas of international priority: Advanced Study Institute, Collaborative Research Grants, Advanced Research Workshop, Networking Infrastructure Grants, Linkage Grants, Expert Visits, etc. Co-operation in this area is carried out through exchange in the spheres of research expertise and output, contacts between scientists and research institutes, joint projects and expert assistance in the new areas of research. Since 1993 nine Ukrainian participants entered into joint ventures together with foreign partners from Germany, Great Britain, France, Norway and other countries.

Ukraine is being integrated into new large scale European Union technological projects through participation in the "Eureka" programme. Together with 11 other countries and East European participants, the Ukraine has been granted the status of an associated member (observer) in the programme. The Ministry of Science and Technology founded the National Information Bureau for Scientific and Managerial Co-ordination of Ukrainian partners. The Bureau has received 1,318 project proposals in 9 technology "Eureka" programme areas. These were examined and passed on to enterprises for expert analysis.

The International Research Fund founded by G. Soros is allotting big sums of money too. In September and October 1994, the Fund signed a long-term research agreement with the Ukraine, Russia and the Baltic States on joint financing of education. The G. Soros financial support programme is very active in the CIS and Baltic States. Among other institutionalised research associations and funds offering financial support to science and education in the Ukraine - European Fund, British Council, International Research Council, IREX, Fullbright programme and others - are markedly active.

The above funds of financial support are primarily focused on competitive sectors of academic, university and specialised areas of research. Of course, this support cannot completely save the local science (Russian scientists approximated that received financial aid forms only about 10 percent of budget expenditures in this field). However, in the situation when the state is in no position to support the research area financially and industry does not commission the product the most active supporter in this situation becomes a foreign organisation. The countries that "pay for the music" dictate developmental priorities to suit their own purposes and with their own aims in mind select areas, best researchers and engineers.

This situation gave rise to the wide spread of the popular misapprehension that these international research institutions and funds are designed to scan local intellectual potential with the aim of attracting the research elite to their countries. Leaving apart a few details, I would like to admit here that Western countries actively help in the S+T area to revive investment and innovation processes and help our scientists to be accepted by the international research community. This situation probably characterises the specificity of science as a very unique and complex social institution without national boundaries destined to serve the interests of all countries and nations. From our point of view it is the science that should become the moving force for political and economic unification of Europe.

Bilateral co-operation between Europe and Ukraine gains more force. In the period between 1992-96 there have been signed more than 26 agreements on intergovernmental and interdepartmental level. With a number of countries we have established governmental commissions for economic, scientific and technological co-operation.

Among European Union members the most active exchange in the area of science and technology is achieved with Germany, France, Denmark and Holland. 31 joint projects with Germany for the total sum of 1.3 million German marks are now in progress.

Scientific and technological co-operation with American Centres and scientists develops somewhat slower than with Western European countries. Here several joint programmes are being run mainly in the area of conversion, as the result of signing the agreement between our governments. Ukrainian scientists participate in Industrial Partnership programmes financed by the US government.

We attach great hopes to the activities of Ukrainian American commission for promoting links between the two States in strategic partnership. The Commission was founded on 19 September 1996 and signed by the President of Ukraine L. Kuchma and Vice President of USA A. Gore. United States are responsible for co-ordination and mobilisation of 1,9 billion US dollars financial aid for Ukraine, the third country US aid recipient. In 1996 Ukraine was given 330 million US dollars in forms of grants and 650 million dollars in credit. Under the auspices of communism, "Kuchma-Gore" there will appear a number of committees for foreign policies, security, trade and investing. The Committee for Permanent Economic Co-operation started its operation.

After disappearance of CMEA in the Countries of Central and Eastern Europe there appeared a certain technological vacuum, also, traditionally existing economic ties between neighbouring countries broke and posed an imperative demand for active political orientation into East European countries. This is also manifest in our attempts to draw nearer to Vyshegrad "square" countries which has been reciprocated by their interest in re-establishing contacts with Ukraine. Governmental commissions for co-operation in economy, science and technology administer contacts in the field of science and technology with Czech Republic and Slovakia also with Baltic States, Lithuania among them. Besides, Ukraine maintains active relations with other regional organisations, for example, with the working group of Central European Initiative and Secretariat of the Black Sea Economic Co-operation.

One of the main priorities for regional exchange is contacts with CIS countries. This is being co-ordinated according to the Treaty of International Co-operation in Science and Technology. 26 international programmes fundamental research projects and 60 scientific and technological programmes were submitted to International Council for Science and Technology since 1992.

In 1995 there were attempts to create a unified scientific zone in CIS which resulted in agreement for creation common scientific and technological zone among CIS countries. In June 1996 first International Science and Technological Co-operation Committee meeting took part in Kiev. The Committee was founded to monitor the above treaty. Scientific and technological integration in CIS develops slowly and in somewhat contradictory manner as opposed to West European situation with open scientific and technological community, common policy and corresponding institutions and mechanisms for integration. The following may be considered as impeding factors: a) widening economic breaches in economies of commonwealth countries; b) deteriorating social and economic situation in the countries and hence diminished compromise possibilities; c) political instability; d) absence of synchronised legal, regulatory, economic and technical institutions and structures in transition period.

And, finally, at present, our main concern is the future of scientific and technological exchange between the Ukraine and European countries. On a very general basis it is possible to predict now that it will depend greatly on the Ukrainian position in the scientific and technological matters.

In other words, it will depend on a sound and flexible doctrine of the Ministry of Science and Technology with regard to ideological needs of economic reform, general tendencies of scientific and technological revolution and development of the world economy, foreign expertise, home and foreign policy priorities, also on certain mistakes and omissions.

All Ukrainian Scientific Development Conference held in February 1996 resulted in some optimistic conclusions. the President of Ukraine, who took part in the forum with his speech "The country without Science Is the Country Without Future," emphasised the necessity of a wide reaching programme for basic reorganisation of co-ordination system in scientific and technological development of the country, including foreign co-operation.

From now on the questions of science development, reorganisation and effective application will be attended personally and controlled by the head of Ukrainian government who becomes the head of the Science Council too. The first step in this direction was founding the Ministry of Science and Technology which responded to the long felt reorganisation needs in monitoring scientific and technological progress. What will the second step be, we shall see in time. However, the importance of the situation lies in the fact that the highest officials in the country are fully aware of the research situation and the need for support for the Ministry of Science and Technology, since we link our hopes of entering the 21st technological century via these institutions.

DYNAMICS OF SCIENTIFIC POTENTIAL OF THE REPUBLIC OF BELORUS AND THE PROBLEM OF BRAIN DRAIN: SHORT TERM AND LONG TERM TRENDS

LEONID ZAIKO,
Institute for Independent Research of the NCSI " East-West"

Social demand for science and scientists exceeded economic expediency in the Soviet times. That was a peculiarity of that type of the socialist society. Further opening of the society and economy in post Soviet countries lead to change of the system of priorities and values, sharp increase of scientific contacts and exchanges. As a result pure hypothetical possibility to work abroad became real for many categories of scientists and specialists.

Simultaneously development of market processes in Belorus showed that certain initiatives and managerial skills could quickly result in "making money" and using personal potential in this sphere faster and more efficient than being a scientist abroad. Attitudes changed at the same time and you can hardly name winners and losers in these processes. Some reservations should be made to understand the situation.

Thus a few young scientists in the Republic became organizers of banks, exchanges, financial and trade companies. Some of them having acquired certain amount of capital got involved into politics or landed up in "big politics" comprising the most perspective and dynamic part of political elite. It is positive in the long run as the society wins at the expense of considerable increase of aggregate intellectual potential of potential candidates for reduced number of posts of history makers. But this factor is not absolute. Growing populism as a method of the political life made possible appearance of other figures were definitely less competitive with new intellectually oriented political elite. This could be called "Belorussian phenomenon".

As for short term changes they can be easily observed first of all in quantitative parameters of scientific institutions and number of scientists and researchers. Slackening transformational processes in the Republic of Belorus went along with steady process of demand-supply regulation in the market of scientific personnel. According to the information of the Institute of Sociology of the Academy of Science of Belorus (ASB) (82 scientific research institutes (SRI), 18 higher educational institutions (HEI) were examined) number of scientific personnel in SRIs decreased by 35 per cent. Number of teachers in HEIs dropped by 4 per cent. We can presume that short term changes happened only at this

K. Prunskienė and E. Altvater (eds.), East-West Scientific Co-operation, 141–144.

period of time. As for long term changes (training personnel of HEIs) we can say that this process lacks elasticity.

Analysis shows that interior migration, transfer of intellectual potential to other spheres of activities took place: either to commercial structures or to non-state HEIs and production enterprises. About 13000 scientific workers changed occupation and there is no information where 65 per cent of them got involved. As for external migration only 293 scientists left the Republic and changed permanent place of residence. Major part of temporary emigrants left the country as they had temporary contracts with foreign organizations (549 persons) and only a small part of them (29 persons) did not come back after expiration of their contracts.

Here is the distribution of the external emigration:

External emmigration of scientists in Belarus (1991- 1995) % to total number

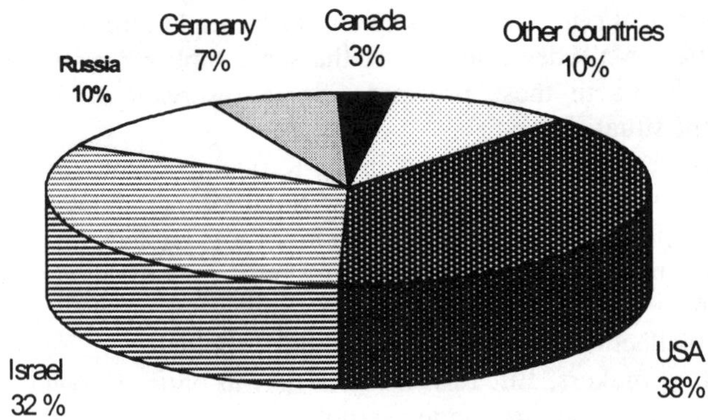

The fact that scientists and teachers left primarily for the USA and Israel enables us to suggest that reasons for their leaving the country were far from being connected with peculiarities of scientific development in their fields of knowledge and research. It is likely to be usual emigration which could result in change of the country, profession and occupation. Only 26 Doctors of Sciences and 114 Ph.D.

scientists (about 3 per cent of overall number of Doctors and 1.5 per cent of Ph.D. scientists).

Thus there was no mass emigration and brain drain. Initial fears of mass emigration turned out vain. The following factors influenced this process: expectations of positive changes, patriotism, inability to quickly adopt to a new place of residence, dynamism of internal changes. It was possible to make "quick" money or an impressive political career in the Republic. The gold mine it were appeared inside the country.

Mobility of scientists is determined not only by demand for scientists and researchers abroad but also by specific character of their work. In respect for this migration indices differ. Residential peculiarities of Belorussian emigrants had the following dynamics:

Dynamics of exernal emmigration
(1991/95 number of persons)

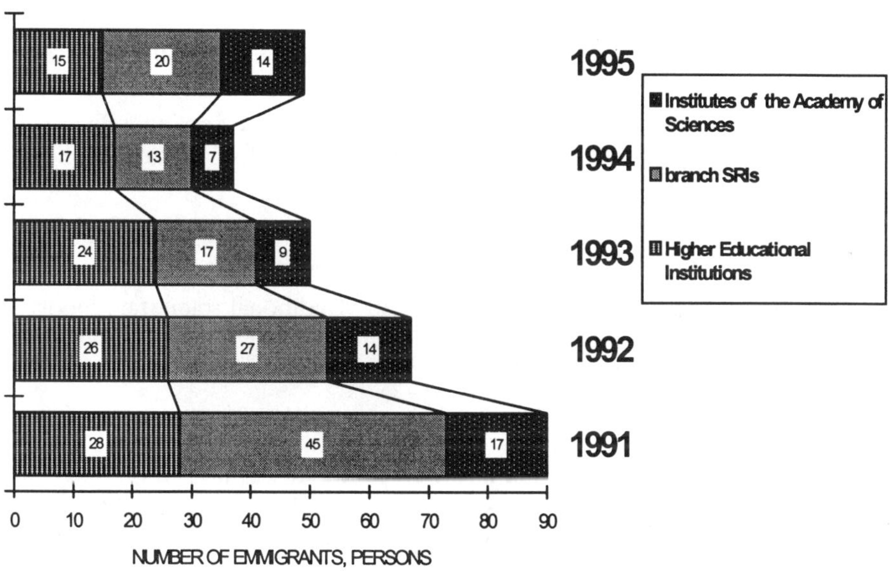

The most numerous group of emigrants consisted of the employees of branch SRIs. HEI teachers were least presented in this group. There are many reasons for that. "Branch scientists" have many more practical applicable projects and demand for owners of such information and patents will always be higher. As for HEI teachers their leaving for teaching at universities and colleges was quite problematic and doubtful due to many reasons: language barrier, level of assumed income and sc on. In fact it is quite difficult to determine the level of life of an immigrant. A certain part of qualified scientific workers are employed as laboratory assistants (this category of workers found jobs quicker than others). Others had to accept odd jobs with no connection to scientific activities.

In the short run development of scientific contacts between the countries of the West and East, change of demand for scientific workers lead to dominance of the following trends:

- sharp decrease (38 per cent in branch SRIs) scientific personnel, development of internal migration;
- accelerated internal migration, increase of intellectual "background" in a number of spheres of activities connected with business and politics;
- growth of temporary departures abroad (conferences, training, workshops) 1974 persons attended such international events held in different countries primarily in Poland, Germany, France, the USA, Italy;

Long term trends in the scientific sphere have the following peculiarities:

- drop of employment in the sphere of academic and applied sciences (in technical sciences number of employed decreased by half, in natural sciences - by a quarter);
- keeping teachers staff of HEIs (though many young people left this sphere);
- growth of persons with scientific degrees;
- restructuring of sciences and weakening of traditional scientific schools and fields of research, loss of leading positions in perspective spheres of knowledge due to financial constraints;
- small scale external migration.

Thus there was not brain drain in the Republic. It was just an ordinary shift of scientific personnel, restructuring of schools and fields of research. In fact that was the beginning of formation of the market of scientific services and corresponding segment of labor on the national and regional labor market.

SCIENCE AND TECHNOLOGY CO-OPERATION: THE ROLE OF THE U.S. OFFICE OF NAVAL RESEARCH EUROPE

STUART J.D. SCHWARTZSTEIN

I represent an organisation whose primary function is in fostering international co-operation in science and technology and that is what I am here to discuss today * not the theoretical aspects, but actual means by which co-operation between scientists is supported and sustained.

Before I go any further, however, I want to thank the cosponsors of this conference, Professor Altvater of the Free University of Berlin and Professor Prunskiene of the Lithuanian-European Institute and commend you for both an extremely interesting conference and some excellent exchanges of views. It is my sense that efforts of this nature are not just important, but critical to improved ties and expanded co-operation, in science and technology, as well as in many other fields. I should also like to thank our Lithuanian hosts, both official and unofficial for their warm hospitality and great helpfulness.

As I think we are all acutely aware, we have now a world that is much changed and, in many significant respects for the better. The cold war is in the past and Lithuania and the Baltic states are now, again independent. With the sad exceptions of the Balkans and Chechnya, war in Europe is also, I believe, a thing of the past; it is my sense that for most Europeans violent conflict in Europe is unthinkable. Most of us now look forward to an era of co-operation and friendly competition, even as we remain aware of dangers and challenges ahead.

My focus, and much of the focus of this conference, of course, is co-operation in science and technology. Over time, both the underlying reasons for co-operation in science and technology and the partners in such co-operation have changed. Co-operation with a wide variety of European states, including those which were part of the Soviet bloc, is now possible and, indeed, very much welcomed by the United States, as by a number of other countries. Co-operation is now, and should be, a matter of choice and it should be of benefit to all partners.

But in addition to the value of scientific co-operation for its own sake * the importance of which should not be in any way minimised * as I think most, if not all, of you know, the United States is certainly strongly interested in seeing the enhancement of bilateral relationships throughout the region, as well as in economic and technological development. The science and technology co-operation we seek is seen as both part of larger co-operative relationships as well as mutually beneficial in specific scientific disciplines. While our role at the Office

K. Prunskienė and E. Altvater (eds.), East-West Scientific Co-operation, 145–148.

of Naval Research Europe (ONREUR) is part of and subordinate to larger frameworks, we think our role is an important one and we hope that our efforts will be of value in advancing basic scientific research.

Our role at the Office of Naval Research Europe is, in essence, to facilitate and make possible a wide variety of co-operative relationships. Our emphasis (and most of our efforts) is on co-operation in basic sciences and fundamental research * as contrast with technological applications. Our broad objectives are to foster the sharing of research results for common usage and to realise the benefits of collaboration in a range of scientific and technical areas among researchers of different nations.

This approach is based on the recognition of three principles: 1) that collaborative work frequently achieves considerably more than independent efforts; and 2) that the a variety of perspectives, talents and approaches can provide an enormous assist in scientific and technological endeavours; and 3) making the most of scientific advances requires communication and access to the best scientific efforts available, wherever they may be found.

In the more than fifty years since ONR Europe was established, there has been a long evolution, from efforts at the inception to assist with the war effort, then a few years later with the post-war reconstruction of Europe, through efforts focused on Western allies' co-operation in the cold war, to the current era in which we see much of our role in fostering co-operation with countries formerly part of the Soviet Union and the Warsaw Pact. ONR Europe is, obviously, a military organisation, but with both a mandate to concentrate on science and technology and the recognition that they cannot be easily categorised as either "military" or "civilian" our focus is very broad. There is, of course, some emphasis on some obvious areas * such as oceanography, for example * in which the Navy has a deep interest. The summary statement we use to describe our mission is that, in the interests of Naval Research, we: "facilitate the exchange of knowledge between the American scientific and technical community and their counterparts throughout Europe, Africa and the Middle East." (I might note that most of our efforts are in Europe, for the simple reason that that's where the science is.)

There are a number of reasons that we believe that it is highly worthwhile to serve such a facilitating function, including:

to capitalise on opportunities for international co-operation to foster advances in science and technology, broadly to improve commonality and interoperability with coalition partners to support programs, policies and requirements of the U.S. Navy to help the Chief of Naval Research and other U.S. Government science and technology managers to formulate policies, programs and budgets * including those that make the U.S. better able to engage in collaborative efforts to establish long-term co-operation in science friends and allies.

Our organisation conducts its work openly, and without secrecy. We have no surveillance functions, do no intelligence gathering and we are engaged in no efforts that cannot be made public. Our work products are published in the open literature * and are even on the Web -- and our offices in London are open to visitors. (I hope to welcome many of you there.)

Most of the research that we support is conducted in universities and research institutes, government and non-government laboratories and industry R & D facilities * wherever we see research that is of scientific value. Our interest is in the best science available.

The means by which we facilitate exchanges of knowledge are varied, but most importantly include:

Fostering liaison of scientists, engineers and technical experts through conferences, workshops, visits and personal research.
Providing assessments of science and technology plans, policies and key activities and accomplishments throughout the region we cover. Exchanging information and ideas in areas of mutual interest to identify opportunities for co-operation

We have several specific programs designed to accomplish this:

Scientific and Technical Liaisons: meetings and site visits for the discussion of areas of mutual interest. The Visiting Scientist Program (VSP): providing support for European scientists to attend science and technology activities (conferences, professional meetings, etc.) in the U.S. This is seen as particularly important in enabling new researchers to establish contacts and develop collaborative proposals, but has also been of value to a number of established researchers and scientists in sharing findings and methodologies with scientific peers. The Conference Support Program (CSP): grants for international conferences and workshops, which provide the opportunity for exchanges of views, findings and information of all kinds in a specific area.
Naval International Co-operative Opportunities in Science and Technology Program (NICOP): providing funds, along with other U.S. Navy cosponsors for international collaboration. This important new programme places emphasis on new science and technology opportunities and new relationships, in order to both develop mutual understanding and build a foundation for further collaborative efforts. Fundamental to this programme is in building on the scientific and technological strengths and capabilities of individual countries. This is a cost-sharing programme (with at least 50% of costs from other sponsors) as well, one in which non-U.S. investigators at institutions outside the U.S. work with U.S. investigators.
Adjunct Scientist and Engineer Program: working with other scientists and engineers to better understand scientific developments throughout Europe

Most of the products of our work are in the form of reports, newsletters, data bases and now, in co-operation with Army and Air Force counterparts in Edison House, an Internet "Home Page" (http://www.ehis.navy.mil/) with a compilation of reports and activities

ONR Europe maintains close contacts with a variety of government and non-government science and technology establishments, in the United States and in a wide range of other countries. In the U.S., those include, in addition to the headquarters of the Office of Naval Research, a wide variety of federal government laboratories, industry laboratories, universities and research institutions, as well as such well-known organisations as:

The National Science Foundation
Defence Advanced Research Projects Agency
National Oceanographic and Atmospheric Administration
National Aeronautics and Space Administration

Outside the U.S., we maintain contact with a an equally-broad array of scientific and technical organisations, including universities and research institutions, multilateral science and technology organisations and research centres and industry research groups. These include, for example, a number of European Union S & T organisations, NATO and its Science Programme and the national research centres of a variety of countries.

Let me close with the suggestion that those of you who may be interested in working with the Office of Naval Research Europe, contact us for further information. We can be reached in the following ways:

Internet homepage: http://www.ehis.navy.mil/
e-mail: For details of science programs: mshear@onreur.navy,mil
(my own e-mail: sschwartzstein@onreur.navy.mil)
our address:

ONR Europe
Edison House
223/231 Old Marylebone Road
London NW1
United Kingdom

INDEX